Unpacking Weight Science – The Podcast

Supporting materials for Episodes 1-12

Fiona Willer, AdvAPD

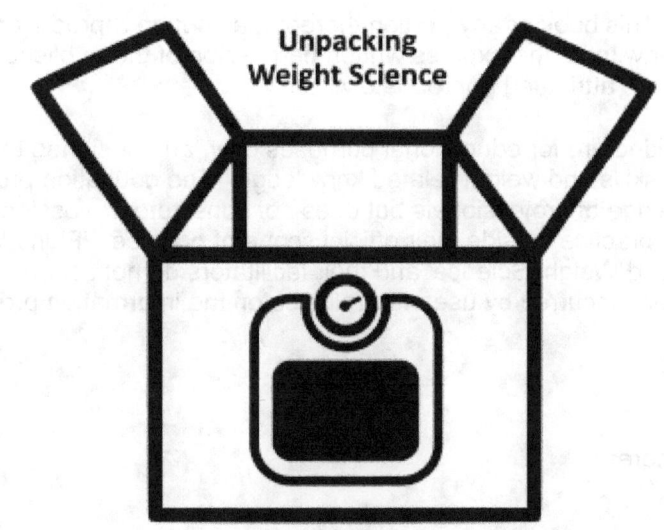

Unpacking
Weight Science

Health, Not Diets
NON-DIET APPROACH™
TRAINING AND WORKSHOPS

First Printing: 2018

Health, Not Diets
PO Box 661, Kenmore
Queensland 4069

Fiona Willer can be reached at fiona@healthnotdiets.com

www.healthnotdiets.com and www.unpackingweightscience.com

Table of Contents

Introduction to the Unpacking Weight Science Podcast

Welcome!

Mainstream weight science is built upon a plethora of assumptions. We now know that many of these assumptions about human body weight and health are flawed. The Unpacking Weight Science Podcast is designed to help you learn how to interpret weight-centric research for yourself and bust myths about body weight and health. It also presents weight neutral research and discusses relevant content (including Health at Every Size®, Intuitive Eating, and The Non-Diet Approach). Because each episode includes further reading and a five question quiz you may be able to claim it as professional development, depending on your credentialing requirements.

I'm a dietitian and academic from Australia, currently undertaking PhD research in weight-neutral approaches to health enhancement and their integration into healthcare practice. My business, **Health, Not Diets** (www.healthnotdiets.com), provides workshops, online courses and resources for health professionals who want to practice from a weight-neutral framework.

While I LOVE presenting, lecturing and facilitating, I also LOVE spreading the word on social media. The nature of these formats means you either get A LOT of content, or A LITTLE content - this podcast is designed as a middle ground (the 'JUST RIGHT' if you will).

Whether you're a health professional, researcher, health science student, getting hassled about your own weight, or just plain interested, this resource will help you to break down what you think you know about weight, and be up-to-date with what we DO know - and it might not be what you think!

Subscribers to the Unpacking Weight Science Podcast (www.patreon.com/UnpackingWeightScience) get a new, 20-30 minute Unpacking Weight Science explainer podcast twice a month, along with these supporting materials, references and quiz. They also get access to the full back catalogue of podcasts, plus the current and previous month's supporting materials. Anyone can subscribe, and it's only $5 a month! After 6 months the podcasts are released to the public and the supporting materials are bundled into.....this!

I hope you enjoy this sojourn into weight science, and I hope it emboldens you to call out the flaws and harms of weight centric science in your own personal and professional life.

I can be found spreading weight inclusive content on Instagram (@FionaWiller), Twitter (@FionaWiller) and Facebook (Health Not Diets).

Warm regards,

Fiona

Episode 1: STUCK IN A WEIGHT CENTRIC OPERATING SYSTEM

Why is it that everyone seems to be pushing a weight-centric agenda? Weight centrism is a cultural norm, medical behemoth, and capitalist bonanza. This podcast delves into the forces that bind us externally and internally, to weight centrism.

Learning Outcomes:

- Recognise the various individual, social and structural factors which currently reinforce weight centrism
- Appreciate some of the ways that weight centrism results in discrimination and unethical healthcare practice
- Appreciate the impact of Confirmation Bias on own beliefs and the beliefs of others
- Recognise the phenomenon of Cognitive Dissonance when presented with information that conflicts with own beliefs

Key Reading:

Confirmation Bias:
Kaptchuk, Ted J. "Effect of interpretive bias on research evidence." *Bmj* 326.7404 (2003): 1453-1455.
https://www.ncbi.nlm.nih.gov/pmc/articles/PMC1126323/

Cognitive Dissonance: (relevant experimental study)
Ciao, Anna C., and Janet D. Latner. "Reducing obesity stigma: the effectiveness of cognitive dissonance and social consensus interventions." *Obesity* 19.9 (2011): 1768-1774.
http://onlinelibrary.wiley.com/doi/10.1038/oby.2011.106/full

Weight-centrism and it's impacts:
Bombak, Andrea. "Obesity, health at every size, and public health policy." *American journal of public health* 104.2 (2014): e60-e67.
https://www.ncbi.nlm.nih.gov/pmc/articles/PMC3935663/

Episode 1 Show Notes:

Arenas where weight centrism is the norm:
- Research: '1001 ways' to lose weight over 2-6 months, follow-up periods usually too short to report full pattern of weight regain even though 12 months is reportable as 'long-term' follow-up.
- News media: 'Obesity epidemic', stigmatising images, advertorials for weight loss products/programs, editorials from anyone with an opinion and a weight loss idea/business
- Social media and advertising: weight loss stories, product promotion, aesthetic advertising, hideous comments on any post or article supporting size acceptance
- Clinical practice: broad 5-10% weight loss recommendation in clinical guidelines, endless referrals for 'weight loss', limited recognition of informed consent in weight-related approaches
- Professional training: prepares us for weight centric practice, prioritises obesity as a nutrition problem rather than knowledge/behaviour/psychological issues
- Professional identity: 'experts in weight management'
- Healthcare: any treatments delayed based on BMI (IVF, joint replacements), providers are frequently organizationally obligated to record and discuss weight

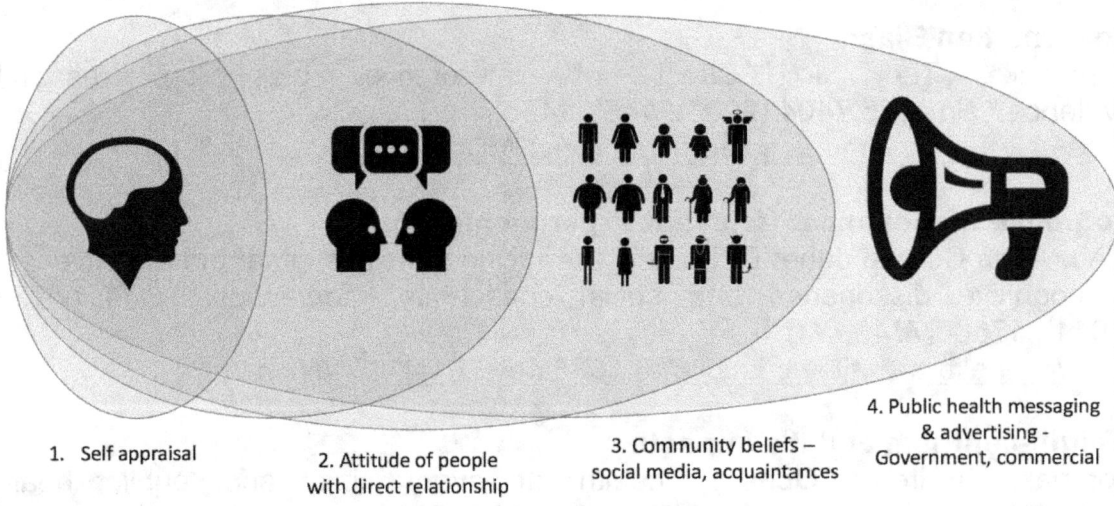

1. Self appraisal

2. Attitude of people with direct relationship

3. Community beliefs – social media, acquaintances

4. Public health messaging & advertising - Government, commercial

I'll cover BMI in more detail in a few weeks, including it's history, but for our purpose today, the summary is – a relationship between higher body weight and younger age at death than predicted was noticed. Later it was noted that some chronic diseases had an earlier onset and greater prevalence in people in larger bodies when compared with people in smaller bodies. These relationships are the basis of all of the medical and research concern regarding body weight. It was (understandably but incorrectly) thought that if someone in a body classified as obese would lose weight to not be obese any more, that their health risks would be the same as someone whose body had always had a lower weight.

It's an example of 'body as a cup of water' thinking – just pour some out, and it'll be indistinguishable from a cup that has been always half-full. Of course, in reality, bodies that have been larger and then become smaller bear evidence of that change, in their body composition, in their metabolic processes, in their bone density, in and on their skin and body tissue, and in the thoughts, habits and experiences of the person to whom the body belongs.

However, that kind of 'cup of water' thinking is still widely prevalent in public health messages – that 'obesity' is bad because it coincides with a range of bad things and everything would be better for everyone if people were no longer obese. Obesity is framed as a problem from a health burden, medical care cost and economic losses perspective. So clinicians, researchers and entrepreneurs have taken that brief – the concept that larger people must strive to lose weight and smaller people must stop themselves from gaining weight – and run with it. The research world abounds with 1001 ways to lose weight over 3-6 months with 12 month follow up studies. The prevailing attitude is that the research team who discovers 'the' solution to obesity will make headlines for eternity, 'save the world', guarantee them research funding forever etc – it's a heromaker. The 'needle in a haystack' thinking is so deeply ingrained that no one dares to consider that there is no needle, and that the vast majority of haystacks have never had needles in them!

For clinicians, our clinical guidelines across the world now recommend 5-10% weight loss for people classified as obese by BMI, and many primary care providers are compelled by their professional guidelines to discuss weight at every consultation with a larger person. The sentiment that a healthy lifestyle has benefits without weight loss appears in every set of guidelines, but is usually at the end of a general paragraph, buried somewhere, unindexed, in the middle of the guidelines. Little attention is given to promoting an informed consent procedure that allows a client to opt out of weight loss discussions all together, or to choose lifestyle changes over weight loss goals. As it stands right now, the number on the scale is considered a, if not the, key health target. A means by which a person can be labelled 'healthy' or 'unhealthy', 'okay' or 'problematic'. This means that people attending their general practitioner for an earache, or unusual bleeding, or a skin check, are supposed to receive encouragement to lose weight. For people who have been trying damn hard to do just that for decades, and/or have a problematic relationship with their body or eating, these unsolicited discussions sting, and drive people away from their providers.

If you feel uncomfortable about some of the concepts discussed in Unpacking Weight Science, I'd like to invite you to examine your thoughts and feelings, before dismissing the concept. Dismissal of an idea because cognitive dissonance feels yucky is an unconscious reaction - to truly examine it you need to go back and give it a do-over. It's a rich opportunity for reflection and growth - in my experience, the stronger the negative emotional reaction someone has to weight-neutral concepts and unpacking of weight science, the more values based their work is, and the more internalised weight stigma and/or fatphobia they hold. Remembering that a belief is a bit like a fact, while an emotion is a reaction to something, an embodied feeling - it is essentially these strong emotions that are pinning their weight centric beliefs to the wall. For an academic with skin in the 'obesity research' game you might first feel horrified, amused, angry, (I've seen all of these reactions!), clinicians can feel all of these things as well as existential angst when they realise that letting go of weight centrism means that a large chunk of their professional work and/or professional identity may disappear. For someone who has had a lived experience of years of trying to reduce their weight, being harassed by loved ones and medical professionals to do so, having repeated weight regains, perhaps usage of weight loss medications or even surgery, considering the world from a weight neutral perspective could be scary as hell or utterly inconceivable. When all meaning about a body has boiled it down to being good or bad due to it's weight or shape, removing that safety blanket can feel like freefall. To those who have not personally been persecuted because of their own body weight or shape, or have not tried unsuccessfully to change it, suggesting that the largeness of a body is unimportant can sound wildly irresponsible, and may produce feelings of anger and frustration.

All of those emotional responses are valid and understandable, given our weight-centric operating environment. But it is the underlying assumptions that these values are based on that is incorrect - evaluate the facts and reform your beliefs, and not only does that yucky cognitive dissonance go away, but you're now armed with the tools to help others find relief from weight-based suffering - just not in the way that you might have originally thought!

Episode 2: DEMYSTIFYING DEFINITIONS, DE-MYTH-DEFYING ASSUMPTIONS

This episode discusses the two main myths that perpetuate weight loss encouragement as well as the terms, weight centric, weight neutral, lifestyle intervention and concern troll.

Learning Outcomes:

- Recognise the key false assumptions that enable the perpetuation of weight-centric attitudes
- Recognise the various common terms used in the size acceptance community, including 'weight neutral', 'weight centric', lifestyle intervention and weight neutral interventions and finally 'concern troll'

Key Reading:

Dietary Quality and BMI:
Guo, X., et al. "Healthy eating index and obesity." European journal of clinical nutrition 58.12 (2004): 1580.
https://www.nature.com/articles/1601989

Weight centrism vs weight neutrality:
Tylka, Tracy L., et al. "The weight-inclusive versus weight-normative approach to health: Evaluating the evidence for prioritizing well-being over weight loss." *Journal of Obesity*2014 (2014).
https://www.hindawi.com/journals/jobe/2014/983495/abs/ *(this link will take you to the abstract only, but if you search for the article by title in Google Scholar there is a link to a PDF version of the full paper which then downloads automatically)*

More on Myths
Bacon, Linda, and Lucy Aphramor. "Weight science: evaluating the evidence for a paradigm shift." *Nutrition journal*10.1 (2011): 9.
https://nutritionj.biomedcentral.com/articles/10.1186/1475-2891-10-9

Episode 2 Show Notes:

Key Myths

1. People in larger bodies eat too much of the wrong stuff and don't get enough exercise and that is what keeps them large and if they would only stop eating that way and start eating 'properly' and moving their body would return to its 'natural' thinner state that it's desperate to get back to.

2. Lasting weight loss is possible if you just stick to your diet plan.
   - Laziness myth (more: http://sciencenordic.com/overweight-people-are-not-lazy-and-dumb)

Concepts

Larger people do not actually eat considerably more than smaller people
Drenowatz, Clemens (11/2015). "Differences in correlates of energy balance in normal weight, overweight and obese adults.". Obesity research & clinical practice(1871-403X), 9 (6), p. 592.

Article re research findings here: https://atlasofscience.org/is-high-energy-intake-driving-weight-gain/

Larger people are not considerably less active than smaller people
http://www.abs.gov.au/AUSSTATS/abs@.nsf/DetailsPage/4364.0.55.0042011-12?OpenDocument

Mentions

The Minnesota Starvation Study:
Info: https://en.wikipedia.org/wiki/Minnesota_Starvation_Experiment
'That' book:
https://books.google.com.au/books/about/The_Great_Starvation_Experiment.html?id=eS-cdiJcnZgC&redir_esc=y

'Lifestyle' Interventions
Can be weight-centric or weight-neutral

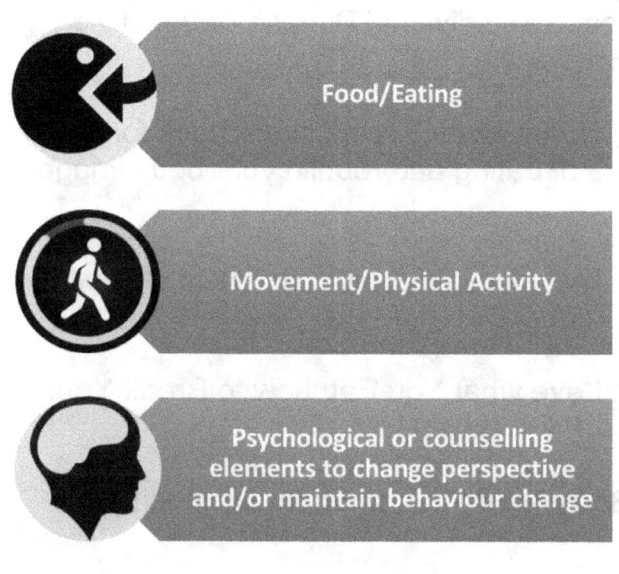

Food/Eating

Movement/Physical Activity

Psychological or counselling elements to change perspective and/or maintain behaviour change

Weight Neutral Lifestyle Interventions

- **Sensitive to considerations of body size**
  - stigma, discrimination, medical needs
- **Don't aim to change, body weight or shape**
  - Not anti weight loss, simply anti pursuit of weight loss
- **Impact and intention of intervention is not to change weight**
  - Same use of term as in pharmacology

More info: O'Reilly, Caitlin (01/2012). "From Theory to Policy: Reducing Harms Associated with the Weight-Centered Health Paradigm". *Fat studies* (2160-4851), 1 (1), p. 97.

Weight neutral intervention models

For Health Professionals
- Health at Every Size®(HAES®) Principles[17]
- Non-Diet Approach[18]

For public via book or trained coach
- Intuitive Eating[19]
- Well Now[20]
- Am I Hungry[21]
- Body Trust® (Be Nourished)[22]

Underlying motivational framework:
- Self Determination Theory[23]

Counselling styles:
- Motivational Interviewing, Acceptance and Commitment Therapy (ACT)

Body cues & mindful eating

Body/size Acceptance

Nourishment through variety

Self Compassion

Timely, Appropriate, medical care

Enjoyed body movement

Autonomy

References from image:

17 Association for Size Diversity and Health, Health at Every Size Principles, available at www.sizediversityandhealth.org, accessed January 2018.

18 Willer, Fiona L., "The Non-Diet Approach Guidebook for Dietitians: A How-To Guide for Applying the Non-Diet Approach to Individual Dietetic Counselling" 2013, published via Lulu Publishing, Brisbane.

19 Tribole, Evelyn, and Elyse Resch. Intuitive eating: a recovery book for the chronic dieter: rediscover the pleasures of eating and rebuild your body image. 1995.

20 Lucy Aphramor, Well Now Course, http://lucyaphramor.com/dietitian/well-now-course/

21 May, Michelle. Eat what You Love: Love what You Eat: how to Break Your Eat-repent-repeat Cycle. Greenleaf Book Group, 2009.

22 Hilary Kinavey, Dana Sturtevant, Body Trust Program from Be Nourished: https://benourished.org/

23 Ryan, Richard M., and Edward L. Deci. "Self-determination theory and the facilitation of intrinsic motivation, social development, and well-being." American psychologist 55.1 (2000): 68.

Episode 3: HOW WE GOT HERE: BMI MEETS DEATH

A history of the development of the BMI (including how 'obesity' ended up on child growth charts), how we discovered there was a relationship with weight and death, and how a few powerful men gifted us the weight-centrism we have today.

Learning Outcomes:

- Recognise the historical context of the relationship between BMI and health recommendations
- Identify the key players in setting the agenda for weight centrism
- Appreciate the statistical characteristics of childhood growth charts and BMI risk charts

Key Reading:

Anthropometry and Childhood growth charts:
Cole, T. J. "The development of growth references and growth charts." *Annals of human biology* 39.5 (2012): 382-394.
https://www.ncbi.nlm.nih.gov/pmc/articles/PMC3920659/

BMI development and history of categorisation:
Fletcher, Isabel. "Defining an epidemic: the body mass index in British and US obesity research 1960–2000." *Sociology of health & illness* 36.3 (2014): 338-353.
https://onlinelibrary.wiley.com/doi/full/10.1111/1467-9566.12050

Critical review of BMI:
Nuttall, Frank Q. "Body mass index: obesity, BMI, and health: a critical review." *Nutrition today* 50.3 (2015): 117.
https://www.ncbi.nlm.nih.gov/pmc/articles/PMC4890841/

BMI definition

Body Mass Index (BMI) [BMI= height $(m)^2$/weight (kg)] is height-standardised heaviness, a way to compare body weights across different heights. It does not take into account any more detail than height and weight – ie it is not adjusted for muscle mass, age, sex, fitness or any other metric.

Childhood growth

Growth spurts

In this study 5 boys were followed closely across their adolescence – here is how their height and weight spurts happened. They also looked at larger groups of kids – it's fascinating reading!

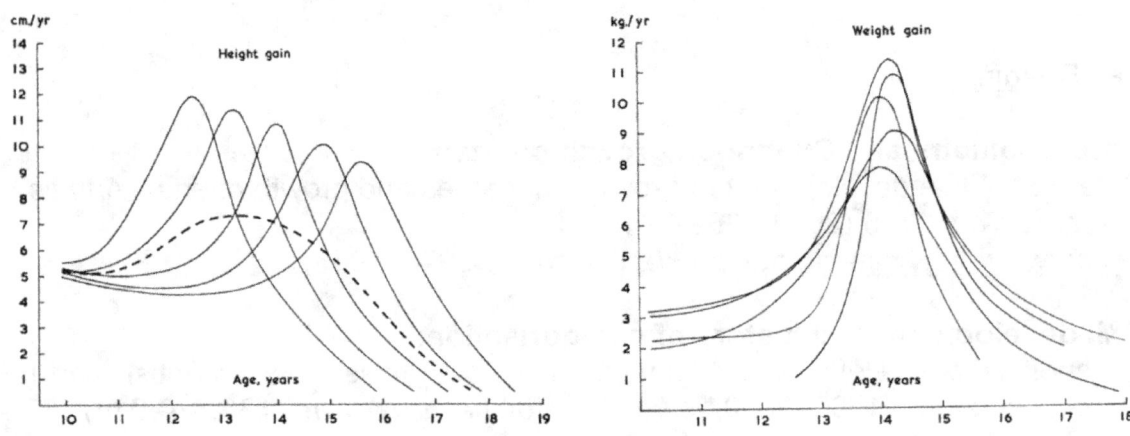

These are taken from the Tanner study: Tanner, James M., R. H. Whitehouse, and M. Takaishi. "Standards from birth to maturity for height, weight, height velocity, and weight velocity: British children, 1965. I." *Archives of Disease in Childhood* 41.219 (1966): 454.
https://www.ncbi.nlm.nih.gov/pmc/articles/PMC2019592/

Changes to terminology

Prior to 1997 growth in childhood was usually expressed in percentiles and the higher BMI kids while seen as a potential 'problem' were not automatically pathologised by BMI alone. In 1997 a consensus panel decided to term the 85th-95th percentile as 'at risk of overweight' and > 95th percentile was called 'overweight'. By 2007 those categories had been renamed 'overweight' and 'obese', shifting the terminology to include the most stigmatising terms. So a 15 year old boy with a BMI of 26 was not seen as problematic in 1987, would have been called 'overweight' by 1997 and labelled as 'obese' by 2007.

BMI for age

Here are the WHO (large cohorts but trying to capture only 'healthy' growth) and CDC (calculated with a smaller but 'real' reference population) charts that depict BMI for age in kids and adolescents:

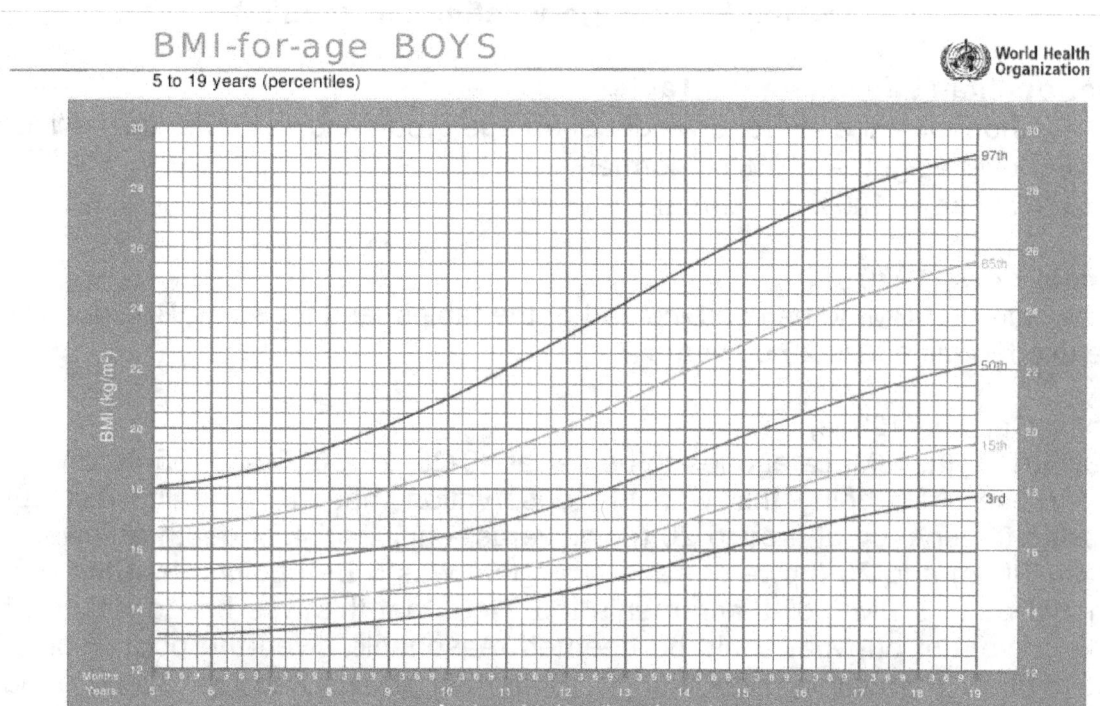

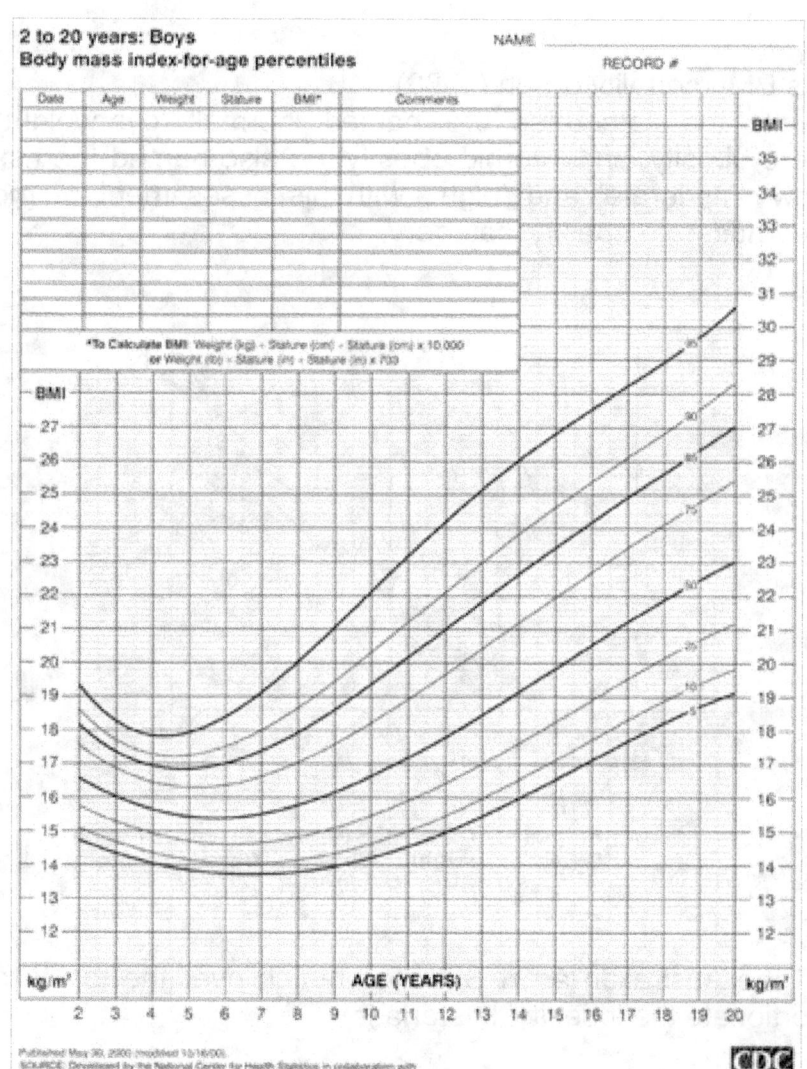

14

Shifting BMI category definitions (adults)

Metropolitan Life Insurance Tables (1959)
- 'normal' = the average weight of an age group and sex, later called 'ideal'
- 'overweight' = 110% of 'normal'
- 'obese' = 120% of 'normal'

Ancel Keys (1970's)
Advocated for BMI to be used instead of the tables, and for same BMI cut offs to be used for all adults 20-65 years

John Garrow (1980s)
Pushed for the BMI categories to be set at 20-25, 25-30, 30-35, 35-40 and 40+. His process for doing this was the observation that the mortality risk curve started to curve up at around 20 (downwards – higher risk of death at lower and lower BMIs) and 25 (upwards) and since that was 5 BMI points, that the others should be in 5 BMI point increments too. Note that this was based only on the risk curve, not on what could be deemed reasonable given the population BMI distribution. Population distribution of BMIs has always had the 'average' adult BMI at at least 25, even during the initial population surveys during the 1960s. He advocated for the 20-25 category to be called 'desirable'.

George Bray
Published this BMI mortality graph (1988)
Note that while the categories are named 'normal' 'overweight' 'moderate obesity' 'severe obesity' and 'morbid obesity' he also named the risk categories 'very low' 'low' 'moderate' and 'high', with his assessment of moderate risk starting at the 'moderate obesity' category.

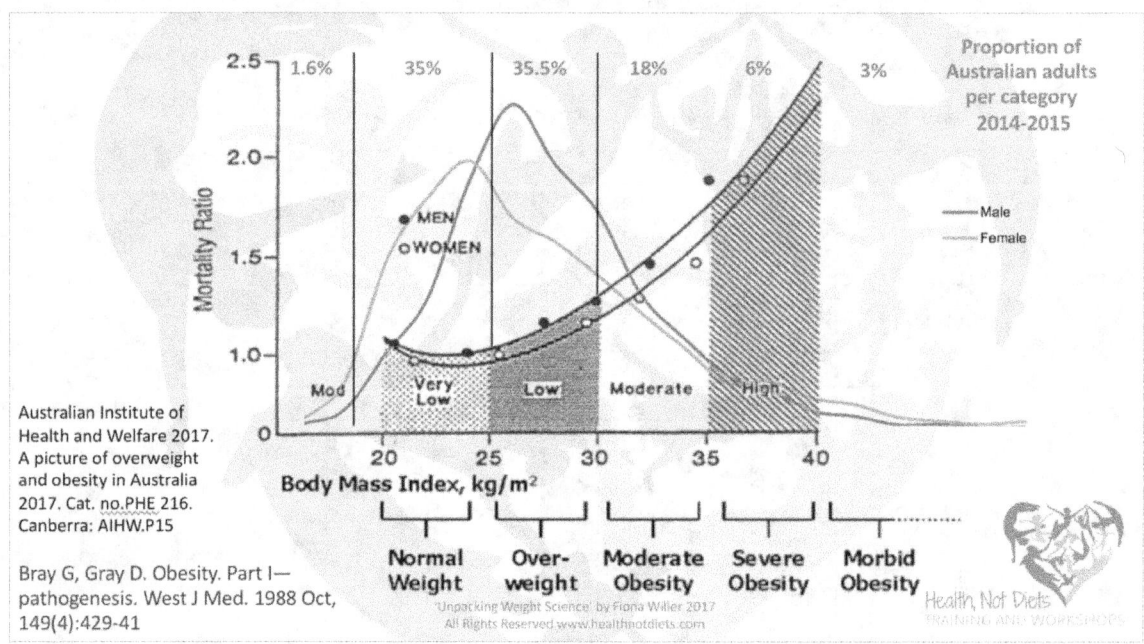

Australian Institute of Health and Welfare 2017. A picture of overweight and obesity in Australia 2017. Cat. no.PHE 216. Canberra: AIHW.P15

Bray G, Gray D. Obesity. Part I—pathogenesis. West J Med. 1988 Oct, 149(4):429-41

(I've overlaid the Australian adult BMI distribution over the top to demonstrate how disproportionate the 'obesity' hysteria is)

US National Institutes of Health (NIH) (1985)

Decided on a cut-off point for 'overweight' in USA to be set at BMI ≥ 27.8 for men and ≥ 27.3 for women. Most British research was already using BMI ≥ 25 for overweight and BMI ≥ 30 for obesity for both sexes.

World Health Organisation (WHO) (1995)

Published the following definitions and BMI cut off points: 'underweight' <20, 'normal' 20-25, 'overweight' 25-30, 'obese class I' 30-35, 'obese class II' 35-40, 'obese class III' 40+

International Obesity Task Force (IOTF) (1997)

Began referring to the 'overweight' category (BMI 25-30) as 'pre-obese', despite evidence showing that most people in that category would never go on to achieve a diagnosis of 'obese'. The trend for 'pre-' diagnoses is referred to as 'pre-disease' and includes 'pre-diabetes', 'pre-hypertension' (interesting paper on this trend here: https://academic.oup.com/epirev/article/33/1/122/487319)

National Institute of Health (and others) (1998)

USA: 'overweight' redefined from 27.8 and 27.3 to >25 to bring it in line with Europe and WHO.

Current academic literature

Now refers to BMI 18.5-25 as 'acceptable', 'normal', 'healthy' and 'ideal'.

Episode 4: WEIGHT BIAS, STIGMA & DISCRIMINATION

This episode is an introduction and defines the terms, discusses their impact on healthcare, society and the lives of the people who are impacted by weight-centrism.

<u>Learning Outcomes:</u>

- Recognise the differences in definitions of weight bias, weight stigma, weight discrimination and weight- based oppression
- Understand the consequences of weight bias for the health system and the individuals who access that health care
- Appreciate strategies that will reduce weight-based stigma and discrimination in a health care setting

<u>Key Reading:</u>

Weight stigma in medical settings
Phelan, Sean M., et al. "Impact of weight bias and stigma on quality of care and outcomes for patients with obesity." Obesity Reviews 16.4 (2015): 319-326.
https://onlinelibrary.wiley.com/doi/full/10.1111/obr.12266

Perceived weight discrimination
Sutin, Angelina R., and Antonio Terracciano. "Perceived weight discrimination and obesity." PloS one 8.7 (2013): e70048.
http://journals.plos.org/plosone/article?id=10.1371/journal.pone.0070048

Societal weight stigma reduction strategies
Puhl, Rebecca M., et al. "Missing the target: including perspectives of women with overweight and obesity to inform stigma-reduction strategies." Obesity science & practice 3.1 (2017): 25-35.
https://onlinelibrary.wiley.com/doi/full/10.1002/osp4.101
Please note that the participants in this study were members of the Obesity Action Coalition (OAC) which has a 'weight loss is good' philosophy BUT they are people living in larger bodies who have experienced weight discrimination so their opinions about weight stigma reduction strategies are valid and useful.

Definitions

Term	Enacted by	Meaning
Weight bias	Individuals, culture/society	Beliefs and preferences *for* or *against* particular body shapes and sizes Can be internalised (pertaining the self *'I don't like my body shape'*) and/or externalised (pertaining others, *'I don't like their body shape'*)
Weight prejudice	Individuals	Inflexible and irrational attitudes and opinions held by members of one weight based group about another *(eg thinner people about larger people, see any weight related article comments section)*
Weight stigma	Experienced by individuals	Negative social impact of weight bias and discrimination *(eg being heckled/verbally abused by strangers, less social 'currency')*
Weight based discrimination	Individuals, groups, organisations, systems	Behaviors (positive or negative) directed against another group, in this case delineated by body size *(eg delaying medical treatments based on BMI, rather than functional assessment)*
Weight based oppression	Groups, organisations, systems	Using systematic power in an unfair and inhumane manner to marginalize people and groups with larger bodies *(eg 'obesity societies' who do not elevate the voices of people classed as 'obese', workplaces who employ only thinner people)*

A person/culture can have weight bias

(their beliefs and preferences *for* or *against* particular body shapes)
Can be internalised (pertaining the self)
and/or externalised (pertaining others)

A person can experience weight stigma

(negative (social) impact of weight bias)

Something that enacts negative bias is 'stigmatising'

(eg a setting a weight-centric policy, making a disparaging comment, sharing a negative belief, using a dehumanising image)

Weight biased focus hijacks client-centred health care

Leads to:
1. **MISSED IDENTIFICATION** of confounders
   - Minimal or no lifestyle assessment, mental health assessment, no screening for disordered eating or body image disturbances (these are instead seen as good/normal)

2. **Patient/client AVOIDANCE** of presenting for health care, sharing struggles, screening, treatment and follow-up
   - Delay in diagnoses, poorer prognosis

3. **OVER-'TREATING'** people who do engage with the health system
   - Lost resources on ineffective treatments
   - 'Lifetime' treatment model for all weight loss interventions

4. **Medication/treatment INCONSISTENCIES**
   - Drugs often not trialed with people in larger bodies – so sometimes less effective or unpredictable results

Impact of weight stigma in healthcare

- People feel **berated** and **disrespected** by providers
- **Upset** by comments about their weight from doctors
- Perceive that they will **not** be **taken seriously**
- Report that their weight is **blamed** for all of their problems
- **Reluctant** to address weight concerns
- Parents of larger children feel blamed and **dismissed**
- Individuals **blamed** for weight status

Anderson, Drew A., and Thomas A. Wadden. "Bariatric surgery patients' views of their physicians' weight-related attitudes and practices." *Obesity* 12.10 (2004): 1587-1595.
Bertakis, Klea D., and Rahman Azari. "The impact of obesity on primary care visits." *Obesity* 13.9 (2005): 1615-1623.
Brown, Ian, et al. "Primary care support for tackling obesity: a qualitative study of the perceptions of obese patients." *Br J Gen Pract* 56.530 (2006): 666-672.
Edmunds, L. D. "Parents' perceptions of health professionals' responses when seeking help for their overweight children." *Family practice* 22.3 (2005): 287-292.
Price, Judy M., and Vid Pecjak. "Obesity and stigma: Important issues in women's health." *Psychology Science* 45 (2003): 6.

Wording for new client materials:
At your initial visit we will ask you if you would like to talk about your weight. This is a weight inclusive practice where we encourage health enhancement through health related behaviours without the expectation of weight loss. If you would prefer not to discuss your weight that will be marked on your file and we won't mention it again unless you do.

What you can do to help:
- Listen to your patients/clients experiences relating to their weight and actively seek their feedback and ideas
- Use health behaviour, function and biochemical screening without pre-empting judgment on the basis of BMI, for people of all weights.
- Focus on disease management without weight loss as a goal, instead focusing on self-determined improvements in eating and activity patterns, informed by scientific research findings.
- Have equipment and clinic set-up which is appropriate for larger bodies and set as the norm, including chairs, examination tables, specula, blood pressure cuffs etc
- Share the evidence of long term ineffectiveness of intentional weight loss and the benefits of size acceptance and health promoting behaviours with your patients/clients
- Remove stigmatising elements from your practice: posters, flyers, magazines which demonise larger bodies, cease policies which discriminate arbitrarily based on size and search for treatments shown to be superior for larger bodies. Display images of people in diverse bodies doing healthy and normal things
- Put pressure on pharmaceutical reps to provide efficacy and safety information for their medications when used by people in larger bodies.
- Work with a multidisciplinary team where indicated, all of whom take a weight neutral approach
- Advertise your services as a size acceptance, weight neutral or Health at Every Size clinician.

Episode 5: THE ANATOMY OF A WEIGHT LOSS PAPER

A research paper is a piece of persuasive writing. This episode breaks down how all the pieces fit together, the expected rhetoric of weight loss research, common research methods that guarantee statistical significance at the cost of clinical practicalities, how to spot hyperbole, and how the way we usually read research papers leaves us at the mercy of the authors biases.

Learning Outcomes:

- Awareness of how to interrogate the methods section of a weight loss paper for non-weight related relevant information
- Appreciate the research paper genre as one in which authors, editors and reviewers have creative influence over the findings narrative, in the conducting of the research to start with and during the publishing process

Key Reading:

Weight loss study claims
Aphramor L. Validity of claims made in weight management research: a narrative review of dietetic articles. Nutrition Journal. 2010;9:30. doi:10.1186/1475-2891-9-30.
https://www.ncbi.nlm.nih.gov/pmc/articles/PMC2916886/

Weight loss maintenance review
Turk MW, Yang K, Hravnak M, Sereika SM, Ewing LJ, Burke LE. Randomized Clinical Trials of Weight-Loss Maintenance: A Review. The Journal of cardiovascular nursing. 2009;24(1):58-80. doi:10.1097/01.JCN.0000317471.58048.32.
https://www.ncbi.nlm.nih.gov/pmc/articles/PMC2676575/
This is a weight centric paper but is a decent summary of types of weight loss studies and the limitations that they have, as well as the follow up times, attrition and small % maintained.

How to critique a research paper
Coughlan, Michael, Patricia Cronin, and Frances Ryan. "Step-by-step guide to critiquing research. Part 1: quantitative research." British journal of nursing 16.11 (2007): 658-663.
https://www.unm.edu/~unmvclib/cascade/handouts/critiquingresearchpart1.pdf
This is a general guide, not specific to weight loss research but is a good overview of the purpose of each section, some study designs and what to look for.

What to look for

- Best intervention studies to take seriously are RCTs which sample their population of interest appropriately, have low attrition, look at health behaviours and disease/event endpoints, have long follow up (preferably until death) and use Intention To Treat (ITT) analysis.
  - ie studies conducted as part of big national cohorts like those in Sweden, Framingham, Nurses study etc. Smaller studies with short follow up times are fairly useless yet continue to receive funding.
- Checklist
  - What did the **control group** do? (what did the intervention **actually consist of** when compared with the intervention group?)
  - **How long** were they followed for? (less than 2 years follow up is SHORT TERM)
  - If more than 20% **dropped out**, why? (hard to follow, odd times, made them feel bad?)
  - Are they comparing **groups at same time point** or **individuals at different time points**?
  - How are they measuring **compliance** with the program? What happens to the non-compliant participant's data?
  - Did they **carry the last measurement forward** to the end if the person dropped out? (danger, danger)
  - Were there any **primary outcomes** other than weight or BMI?
  - Did they measure **surrogate markers** or actual disease/events?
  - Did the study control for **dietary quality, fitness, psychological factors** etc?
  - Who **funded** the study? Was it conducted at a university?

Health, Not Diets
TRAINING AND WORKSHOPS

'Not quite honest' weight loss research factors

- Motivated volunteers to start with (ie **not representative**), sometimes exclusion criteria leaves very narrow group (ie not representative) yet results used by health professionals to justify weight loss efforts for many
- **Short follow up times** so only the weight loss (and biochem improvement) and not the regain (and biochem recidivism) is captured.
- **Last result/observation carried forward** – participant dropped out but their last recorded result (ie weight) is carried forward to the end of the study even though it is unlikely to have remained the same in real life.
- **Per Protocol (PP) analysis** removes those deemed non-compliant from analysis, and those who dropped out or had missing data – these are the 'most perfect' participants' but are not then representative – they reflect the 'ideal' outcome rather than the norm.
- Low and high energy intakes are sometimes labelled **'implausible'** and removed from the study – this assumes people are lying – what if they're not??

Health, Not Diets
TRAINING AND WORKSHOPS

Within group analysis vs between group analysis

Differences between the average results
of the intervention group vs the control group

Sometimes both are used, just have your
wits about you when you're reading the study!

Individual differences across time
in the intervention group

Health, Not Diets
TRAINING AND WORKSHOPS

Biochemical Changes

When losing weight, biochemical markers that we flag as risk factors change as a biological adaptation. Most studies of the benefits of weight loss only follow participants for 3-6 months. The temporary suppression of risk factors does not necessarily confer actual disease incidence reduction.

*Use of **surrogate markers** not actual illness end points leads to potential **overestimation** of benefits*

Surrogate marker (risk factor, not independently life limiting)	Permanent damage or life threatening event/condition
High LDL cholesterol (hypercholesterolemia)	Coronary artery disease Heart attack
High blood pressure (hypertension)	Stroke
High HbA1c (% glucose on red blood cells, measure of chronic high blood sugar levels)	Nephropathy, Retinopathy, Neuropathy

Remember – people with low blood pressure have strokes, people with low blood cholesterol have heart attacks, BP and cholesterol are risk factors

Health, Not Diets
TRAINING AND WORKSHOPS

When to put the study in the bin:

- Follow up time 12 months or less
- Intervention required a lot of time, effort or expense
- Any intervention that proposes or necessitates a 'lifetime model of care' after the intervention, beyond regular primary care (eg from your GP at usual visits) is a huge drain on personal and healthcare resources – this is essentially the cost of pathologizing body weight.
- A study that uses last measurement carried forward for any time points that they are missing data for
- Small numbers of participants (50 or less) where the paper is trying to convince readers of an exciting brand new thing
- No information on what the control group was doing, if they used a control group

Episode 6: THE SCIENCE OF SELF COMPASSION

In this podcast I take you through the research and academic operationalisation of self-compassion, some of its associations and effects on individuals and their self-care as well as on health practitioners with compassion fatigue and burnout. I talk about how to weave language that models a self-compassionate attitude into your counselling and encounters with others and yourself. These supplementary materials contain exercises shown in experimental studies to induce self-compassion.

Learning Outcomes:

- Recognise the operational constructs of self-compassion, as defined by Neff
- Appreciate the health factors and outcomes associated with self-compassion
- Identify strategies which have been shown to induce self-compassion in experimental studies

Key reading:

Self-compassion and its relationship with mindfulness
Neff, Kristin D., and Katie A. Dahm. "Self-compassion: What it is, what it does, and how it relates to mindfulness." Handbook of mindfulness and self-regulation. Springer, New York, NY, 2015. 121-137.
http://self-compassion.org/wp-content/uploads/publications/Mindfulness_and_SC_chapter_in_press.pdf

TEST YOURSELF: Self Compassion Scale
Neff, K. D. (2003). Development and validation of a scale to measure self-compassion. Self and Identity, 2, 223-250.
http://self-compassion.org/test-how-self-compassionate-you-are/

More about the Self Compassion Scale
Neff, Kristin D. "The self-compassion scale is a valid and theoretically coherent measure of self-compassion." Mindfulness 7.1 (2016): 264-274.
http://self-compassion.org/wp-content/uploads/2015/12/ScaleMindfulness.pdf

Self-compassion definition

What is self compassion? (Neff 2003)

- Differs from self esteem (which calls for comparison with others)

Components	Description
Kindness	Responding to difficult times or difficult emotional states with a spirit of kindness, warmth and love. Seeking to understand the situation rather than judging it harshly. *I am more than this experience and have many great things to offer. I did my best.*
Common Humanity	Recognising that pain and imperfection are part of the human experience, a normal part of being alive. Seeking to connect to that sense of the larger human experience when times are tough, rather than feeling isolated and alone in your pain. *Feeling this way is really common; everybody has probably felt this way at some point.*
Mindfulness	Observing the internal landscape of thoughts and feelings without becoming overly involved in them. *I am feeling frustrated and sad.*

Self-compassion and health enhancing behaviours

- More realistic and intrinsically motivated exercise goals (Magnus et al 2010)
- More likely to seek medical care quickly (Terry et al 2013)
- Reduces negative affective states (Leary et al 2007)
- Improves positive affective states (Neff 2003, 2007)
- Smoking reduction (Kelly et al 2010)
- Reduced alcohol misuse (Brooks et al 2012)
- Less risky sexual behavior in people with HIV/AIDS (Rose et al 2014)
- Proactive attitude towards health, benevolent self talk, motivation towards self-kindness (Terry et al 2013)

Self-compassion and eating behaviours

- Less negative reaction to diet-breaking scenario in restrained eaters (Adams and Leary 2007)
- Fewer binge eating symptoms (Webb and Foreman 2013)
- Decreased social physique anxiety (Magnus et al 2010)
- Fewer body image concerns after controlling for self esteem (Wasylkiw et al 2012)
- Lower self compassion associated with higher eating disorder pathology in ED patients (Kelly et al 2013)
- Improvement in shape and weight concerns (Albertson 2012)
- High self compassion associated with low disordered eating behaviours (Geller et al 2015)
- May moderate the relationship between distress and disordered eating (Geller et al 2015)

Self Compassion Scale Subscales

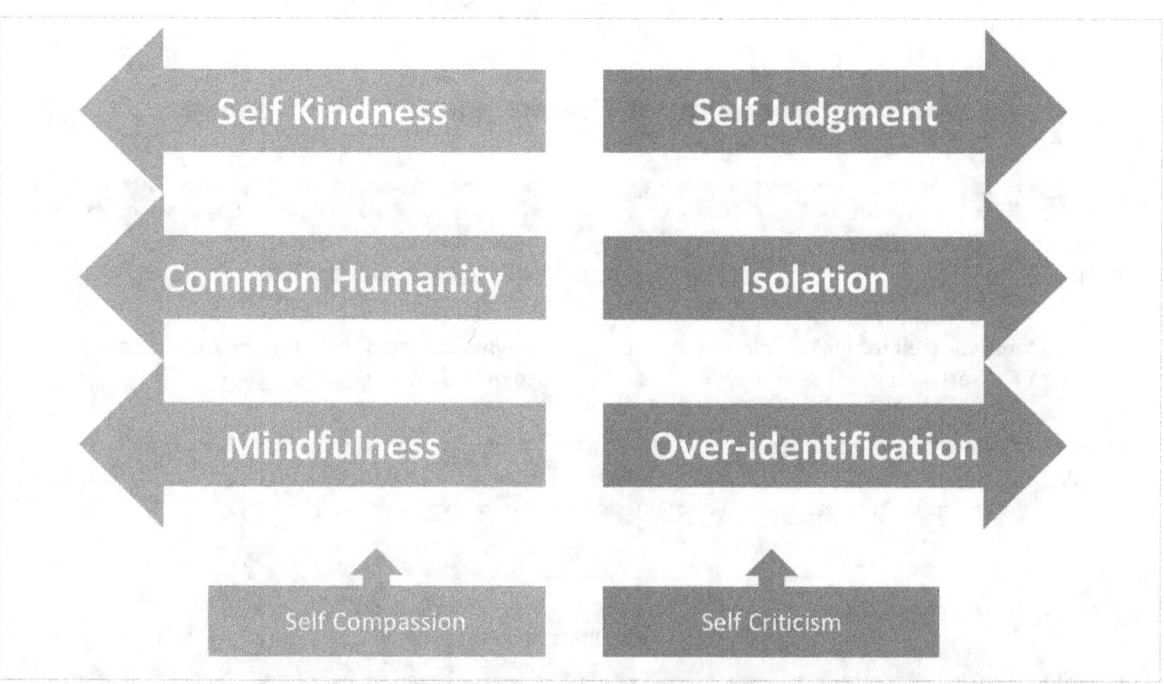

Self Compassion Scale Subscales – examples

	Self Judgment	Isolation	Over-identification
Self Criticism	*I'm such a loser*	*No-one knows how this feels*	*I'm never going to succeed*
	Self Kindness	**Common Humanity**	**Mindfulness**
Self Compassion	*I am more than this experience and have many great things to offer. I did my best.*	*Feeling this way is really common; everybody has probably felt this way at some point.*	*I am feeling frustrated and sad*

Cognitive Behavioural Therapy versus Mindfulness

Typical reaction (and CBT)	Mindfulness (and ACT and DBT)
Challenges and labels disruptive thoughts/emotions	Non-judgemental observation of disruptive thoughts/emotions – creates distance. Disengages from unhelpful thoughts
Trying to force yourself to think a different way (cognitive restructuring)	Noticing thoughts and emotions from another place (inner self) (metacognitive awareness)
Assumption of healthy normality (pain and suffering are pathological (wrong/sick/bad) factors that need to be alleviated or avoided)	Human suffering is a natural part of the human experience. To expect to avoid all suffering is unrealistic. Suffering is a basic fact of human existence.
Strategies to avoid pain	Learn to accept, tolerate and appreciate emotions/thoughts
Behavioural goals set according to given plan	Self-directed, experimental, experiential learning leading to behaviour change

Activity: Self Compassion Induction

(modelled after Leary et al 2007 and Kramer 2014)

This exercise offers you the opportunity for increasing your self compassion. First we induce some pain, and then practice the responses that encourage kindness, common humanity and mindfulness.

1. First you need to 'call up' some past pain/suffering
- Think of a negative life event in your past where you have experienced emotional pain. It could be a time when you made a mistake and that made you feel embarrassed, sad, etc. Do not start with big stuff, choose something painful but not excruciating.
- Set a timer for 5 minutes and start the timer
- Write a description of what happened and how you felt.
- If you finish early, continue to read over what you have written until the time is up.

2. Induce self-kindness
- Imagine that a friend has just gone through or is experiencing what you wrote about in step 1.
- Write a kind letter to your friend responding to their situation. Use detail.
- If it is easier to find compassion for a child or pet, you can use one of those examples instead.

3. Induce an understanding and sense of common humanity
- Can you recall isolating yourself during your painful time? Reflect on whether your world 'shrunk' while you were coping and upon any feelings of isolation from your friends and family during this time.
- Now consider who else might have had similar feelings and experiences to yours. This could be people you actually know, or people who experience similar situations.
- Write a list of the people or types of people who might have had a similar experience to yours, think about what you have in common with them.

4. Conduct mindfulness practice of observing thoughts and feelings
- Write again about the painful experience from step 1 and this time discuss the emotion and thoughts in a neutral and non-judgmental way – writing in third person can help. 'she was experiencing the feelings of sadness and anger..... etc'

Episode 7: WEIGHT-NEUTRAL HEALTH-ENHANCING HABITS

This episode looks at behaviours associated with longevity, delay and avoidance of disease, as well as behaviours which enhance health outcomes at any weight and health status. Key research papers are discussed as well as strategies for how to use them in your work.

Learning Outcomes:

- Appreciate the different definitions and indications of health
- Identify intentional behaviours which are associated with enhanced health indicators
- Identify the types of scientific studies which can shed light on weight neutral health enhancing lifestyle factors

Key reading:

Population BMI, health habits and all-cause mortality
Matheson, Eric M., Dana E. King, and Charles J. Everett. "Healthy lifestyle habits and mortality in overweight and obese individuals." The Journal of the American Board of Family Medicine 25.1 (2012): 9-15.
http://www.jabfm.org/content/25/1/9.full

Fruit, Vegetables and mortality
Aune, Dagfinn, et al. "Fruit and vegetable intake and the risk of cardiovascular disease, total cancer and all-cause mortality—a systematic review and dose-response meta-analysis of prospective studies." International Journal of Epidemiology 46.3 (2017): 1029-1056.
https://academic.oup.com/ije/article/46/3/1029/3039477

The risk of relying on surrogate markers
Fleming, Thomas R., and David L. DeMets. "Surrogate end points in clinical trials: are we being misled?." Annals of internal medicine 125.7 (1996): 605-613.
http://www.tig.org.za/pdf-files/affidavit-aug06/75%20Fleming%20and%20De%20Mets.pdf

Surrogate Markers

Biochemical Changes

Fiona Willer for www.healthnotdiets.com

When losing weight, biochemical markers that we flag as risk factors change as a biological adaptation. Most studies of the benefits of weight loss only follow participants for 3-6 months. The temporary suppression of risk factors does not necessarily confer actual disease incidence reduction.

*Use of **surrogate markers** not actual illness end points leads to potential **overestimation** of benefits*

Surrogate marker (risk factor, not independently life limiting)	Permanent damage or life threatening event/condition
High LDL cholesterol (hypercholesterolemia)	Coronary artery disease, Heart attack
High blood pressure (hypertension)	Stroke
High HbA1c (% glucose on red blood cells, measure of chronic high blood sugar levels)	Nephropathy, Retinopathy, Neuropathy

Remember – people with low blood pressure have strokes, people with low blood cholesterol have heart attacks; high blood pressure and cholesterol are only risk factors, not definitive predictors

Habits associated with longevity

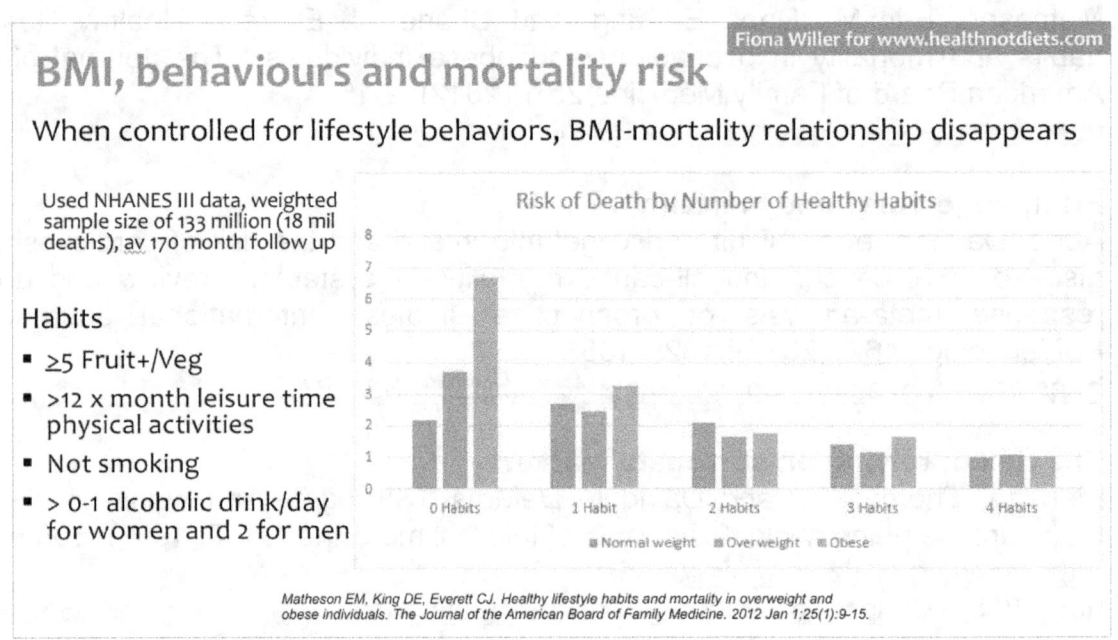

BMI, behaviours and mortality risk

Fiona Willer for www.healthnotdiets.com

When controlled for lifestyle behaviors, BMI-mortality relationship disappears

Used NHANES III data, weighted sample size of 133 million (18 mil deaths), av 170 month follow up

Habits
- ≥5 Fruit+/Veg
- >12 x month leisure time physical activities
- Not smoking
- > 0-1 alcoholic drink/day for women and 2 for men

Risk of Death by Number of Healthy Habits

■ Normal weight ■ Overweight ■ Obese

Matheson EM, King DE, Everett CJ. Healthy lifestyle habits and mortality in overweight and obese individuals. The Journal of the American Board of Family Medicine. 2012 Jan 1;25(1):9-15.

Weight loss is not necessary to improve physical health

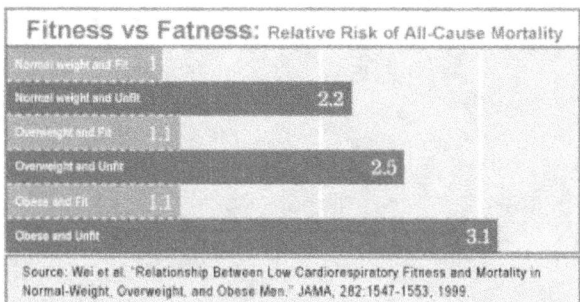

Fitness vs Fatness: Relative Risk of All-Cause Mortality

Normal weight and Fit	1
Normal weight and Unfit	2.2
Overweight and Fit	1.1
Overweight and Unfit	2.5
Obese and Fit	1.1
Obese and Unfit	3.1

Source: Wei et al. "Relationship Between Low Cardiorespiratory Fitness and Mortality in Normal-Weight, Overweight, and Obese Men." JAMA, 282:1547-1553, 1999.

- Studies that have actually controlled for fitness have found that it is more predictive for mortality than weight.

- This study defined 'fit' as 3-4 hrs/week of walking

'Good food' and weight

Assumption: "eat well and your weight will 'normalize' or 'correct itself'"

21% reduced risk of all cause mortality

In a free-living adult population, diet quality, and specifically fruit and vegetable intake did not vary significantly with BMI, but it was associated with a longer lifespan[14]

Thus, habitual consumption of health promoting food does not lead automatically to weight loss

AND

BMI does not accurately reflect eating pattern

Russell J, Flood V, Rochtchina E, Gopinath B, Allman-Farinelli M, Bauman A, Mitchell P. Adherence to dietary guidelines and 15-year risk of all-cause mortality. British Journal of Nutrition. 2013 Feb 14;109(03):547-55.

Dietary quality quintiles in representative sample of Australian adults >49 yrs

(bar chart, x-axis: Poorest quality ... Highest quality; y-axis 0%–100%)

■ Low BMI < 20 ■ Moderate BMI 20-30 ■ Highest BMI >30

Independent Longevity boosters

- **Regular physical activity**
  - 33% reduction in all cause mortality[1]
- **Varied (high quality) diet**
  - 11-42% reduction in all cause mortality[2]
  - (fruit and veg alone ~5% per daily serve to a max of ~25%)[3]
- **Social support**
  - 11-35% reduction in all cause mortality[4]
- **Being a member of a hobby or community group**
  - 44% reduction in all cause mortality in representative Japanese study[5]
- **Enough sleep (5-8hrs/night)**
  - 12% reduction in all cause mortality compared with <5hrs[6]
  - 30% reduction in all cause mortality compared with >8-9 hrs[6]

[1]Nocon, Marc, et al. "Association of physical activity with all-cause and cardiovascular mortality: a systematic review and meta-analysis." European Journal of Cardiovascular Prevention & Rehabilitation 15.3 (2008): 239-246.
[2]Kurotani K, Akter S, Kashino I, Goto A, Mizoue T, Noda M, Sasazuki S, Sawada N, Tsugane S, Japan Public Health Center based Prospective Study Group. Quality of diet and mortality among Japanese men and women: Japan Public Health Center based prospective study. bmj. 2016 Mar 22;352:i1209.
[3]Wang, Xia, et al. "Fruit and vegetable consumption and mortality from all causes, cardiovascular disease, and cancer: systematic review and dose-response meta-analysis of prospective cohort studies." Bmj 349 (2014): g4490.
[4]Shor, Eran, David J. Roelfs, and Tamar Yogev. "The strength of family ties: A meta-analysis and meta-regression of self-reported social support and mortality." Social Networks 35.4 (2013): 626-638.
[5]Minagawa, Yuka (07/2015). "Active Social Participation and Mortality Risk Among Older People in Japan Results From a Nationally Representative Sample". Research on aging (0164-0275), 37 (5), p. 481
[6]Cappuccio, Francesco P., et al. "Sleep duration and all-cause mortality: a systematic review and meta-analysis of prospective studies." Sleep 33.5 (2010): 585-592.

Episode 8: THEORETICAL VS ACTUAL WEIGHT CHANGE: INDUSTRIALISED WISHFUL THINKING

In this episode I compare and contrast what our mathematical formulas about energy requirements and weight loss say should be possible (ie the theoretical) versus what actually happens in real life. I also talk about the history of this energy requirement belief system, who continues to use it and why.

Learning Outcomes:

- Develop familiarity with the equations used to determine energy requirements, their utility and margins of error
- Understand the historical context of energy deficits to elicit human weight loss
- Recognise the features of literature and sources which use a theoretical basis vs an evidentiary basis to claim weight loss success

Key Reading:

Body weight regulation – explanation & critique of adaptive thermogenesis
Dulloo, Abdul G., et al. "Adaptive thermogenesis in human body weight regulation: more of a concept than a measurable entity?." *Obesity Reviews* 13.S2 (2012): 105-121.
https://doc.rero.ch/record/30575/files/dull_ath.pdf

Comparing energy expenditure and predictive equations of requirements
Ullah, S., R. Arsalani-Zadeh, and J. MacFie. "Accuracy of prediction equations for calculating resting energy expenditure in morbidly obese patients." The Annals of The Royal College of Surgeons of England 94.2 (2012): 129-132.
https://www.ncbi.nlm.nih.gov/pmc/articles/PMC3954136/
This paper is useful as it looks at the requirements of people in larger bodies. However, the participants went on to have weight loss surgery and the article is highly weight centric overall. It is firmly used here for illustrative purposes not as an endorsement.

Dieting and weight regain
Dulloo, Abdul G., Jean Jacquet, and Jean-Pierre Montani. "How dieting makes some fatter: from a perspective of human body composition autoregulation." Proceedings of the Nutrition Society 71.3 (2012): 379-389.
http://doc.rero.ch/record/290645/files/S0029665112000225.pdf

Energy Expenditure

Here is the breakdown of proportion of energy expenditure from various processes. BMR is within RMR.

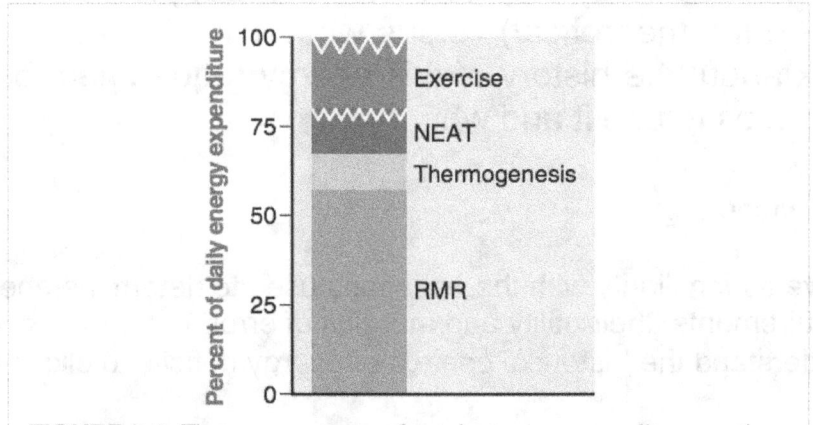

FIGURE 2-2 The components of total energy expenditure: activity, thermic effect of food (TEF), and basal or resting metabolic rate.

From Mahan, L. Kathleen, and Janice L. Raymond. *Krause's Food & the Nutrition Care Process-E-Book*. Elsevier Health Sciences, 2016.

Fluctuations and diversity in virtually everything are normal for free living humans

Here is a graph of weight fluctuations for one person over a 4 year period. Normal weight fluctuations for an individual over 10-30 year periods can be in the magnitude of 7-20% around the average weight. You can see how the fluctuation in a downward direction may resemble intentional weight change but may be entirely attributed to normal human biology.

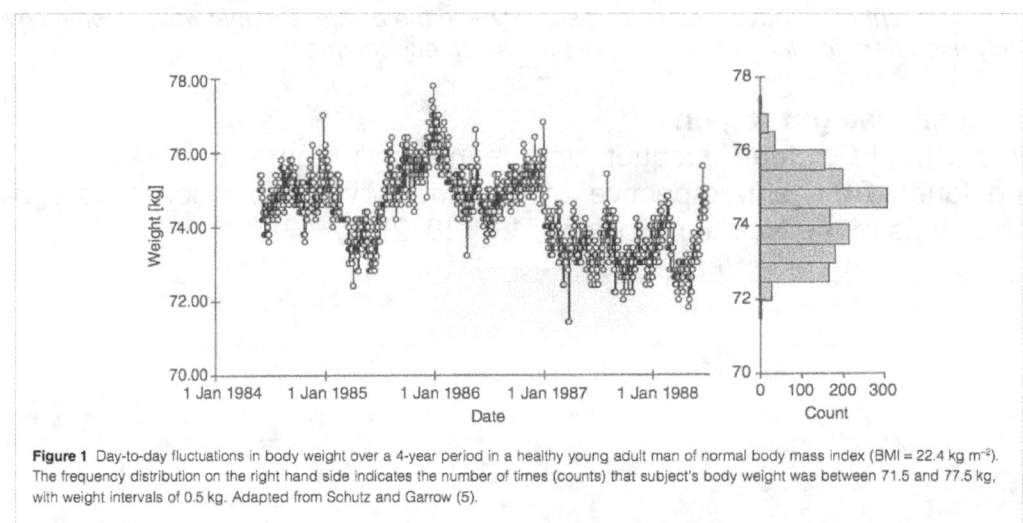

Figure 1 Day-to-day fluctuations in body weight over a 4-year period in a healthy young adult man of normal body mass index (BMI = 22.4 kg m⁻²). The frequency distribution on the right hand side indicates the number of times (counts) that subject's body weight was between 71.5 and 77.5 kg, with weight intervals of 0.5 kg. Adapted from Schutz and Garrow (5).

This is from your first recommended reading.

Variation in energy requirements from person to person

The second recommended reading has a comparison of the different predictive equations compared with the energy expenditure figures that they derived from measuring gas exchange. These participants went on to have weight loss surgery but this table shows their pre-operative energy expenditure:

Demographic data	
Age (years)	47 (SD:7)
Male-to-female ratio	10:21
Height (cm)	168 (SD: 11)
Weight (kg)	134 (SD: 29)
Body mass index (kg/m^2)	46.6 (SD: 8.6)
Resting energy expenditure	
Harris–Benedict equation (kcal/day)	2,195 (SD: 505)
– Men (10)	2,777 (SD: 373)
– Women (21)	1,918 (SD: 266)
Schofield equation (kcal/day)	2,129 (SD: 449)
– Men (10)	2,666 (SD: 307)
– Women (21)	1,874 (SD: 216)
Indirect calorimetry (kcal/day)	1,980 (SD: 558)
– Men (10)	2,329 (SD: 600)
– Women (21)	1,814 (SD: 464)
Overprediction (kcal/day) by	
– Harris–Benedict equation	215 (SD: 458)
– Schofield equation	149 (SD: 439)

SD = standard deviation

You can see that they all had higher BMIs. The energy requirements shown are for RESTING energy expenditure, so this does not take into account the activity factor – their actual energy requirements would be 1.4+ times more than this if they're out living regular lives (ie not bedbound).

The other interesting thing that we can see from this information is the HUGE standard deviations in the indirect calorimetry (breath analysis) section. This is evidence of the wide variation in energy expenditure in their participants. It says that about 70% of their male participants (so 7 men) had resting energy expenditures that fell between 1729 and 2929 calories a day, and the other three must have been spread over a much smaller and much higher amount.

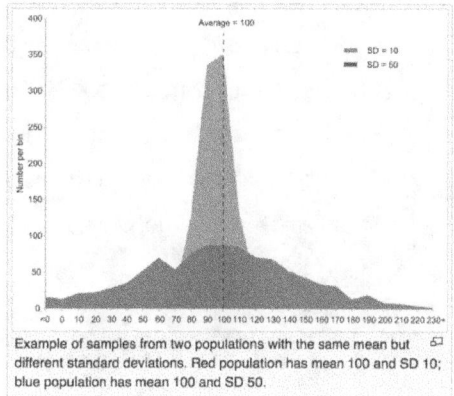

Example of samples from two populations with the same mean but different standard deviations. Red population has mean 100 and SD 10; blue population has mean 100 and SD 50.

Wikipedia has a useful diagram of this
https://en.wikipedia.org/wiki/Standard_deviation
– our results here look more like the blue section.

The way that broad energy intake recommendations are communicated gives the incorrect impression that our requirements are more like the red section.

The average energy intake of 2000 calories fallacy

In Australia the 'reference value' of 8700kj (2000 kcal) is written into our Food Standards Code and the required labelling statement is 'Percentage daily intakes are based on an average adult diet of 8700 kJ. Your daily intakes may be higher or lower depending upon your energy needs.' This is similar for the US and UK labelling requirements. This labelling gives the impression that a) about 2000 calories is a reasonable amount for an average adult and b) that the required adjustments 'depending on your energy needs' are readily available to the public.

In fact, even for most people who have a BMI that is deemed a human 'ideal' (22! not even recognising variation within a range anymore!), that amount of energy is insufficient. I've marked up the table from the Australian Dietary Guidelines (https://www.nrv.gov.au/dietary-energy) showing in yellow the only circumstances where 8700kj would meet requirements. Note that a PAL of 1.2 (the first column) is to use when people are confined to bedrest, and adults who meet the physical activity guidelines would have a PAL of 1.75+

> [b] PAL ranges from 1.2 (bed rest) to 2.2 (very active or heavy occupational work). PALs of 1.75 and above are consistent with good health. PALs below 1.4 are incompatible with moving around freely or earning a living. PALs above 2.5 are difficult to maintain for long periods

Adults

Table 3 - Estimated energy requirements of adult using predicted BMR x PAL

Age yr	BMI = 22.0[a]		BMR MJ/d	Physical activity level (PAL)[b] Males MJ/day						BMR MJ/d	Physical activity level (PAL)[b] Females MJ/day					
	Ht (m)	Wt (kg)	Male	1.2	1.4	1.6	1.8	2.0	2.2	Female	1.2	1.4	1.6	1.8	2.0	2.2
19-30	1.5	49.5	-	-	-	-	-	-	-	5.2	6.1	7.1	8.2	9.2	10.2	11.2
	1.6	56.3	6.4	7.7	9.0	10.3	11.6	12.9	14.2	5.6	6.6	7.7	8.8	9.9	11.1	12.2
	1.7	63.6	6.9	8.3	9.7	11.0	12.4	13.8	15.2	6.0	7.2	8.4	9.6	10.8	12.0	13.2
	1.8	71.3	7.4	8.9	10.3	11.8	13.3	14.8	16.3	6.5	7.7	9.0	10.3	11.6	12.9	14.2
	1.9	79.4	7.9	9.5	11.1	12.6	14.2	15.8	17.4	7.0	8.4	9.7	11.1	12.5	13.9	15.3
	2.0	88.0	8.4	10.1	11.8	13.5	15.2	16.9	18.6	-	-	-	-	-	-	-
31-50	1.5	49.5	-	-	-	-	-	-	-	5.2	6.3	7.3	8.4	9.4	10.4	11.5
	1.6	56.3	6.4	7.6	8.9	10.2	11.4	12.7	14.0	5.5	6.5	7.6	8.7	9.8	10.9	12.0
	1.7	63.6	6.7	8.0	9.4	10.7	12.1	13.4	14.8	5.7	6.8	8.0	9.1	10.3	11.4	12.5
	1.8	71.3	7.1	8.5	9.9	11.3	12.7	14.2	15.6	6.0	7.2	8.3	9.5	10.7	11.9	13.1
	1.9	79.4	7.5	9.0	10.4	11.9	13.4	14.9	16.4	6.2	7.5	8.7	10.0	11.2	12.5	13.7
	2.0	88.0	7.9	9.5	11.0	12.6	14.2	15.8	17.3	-	-	-	-	-	-	-

It is unlikely that no one except the highly motivated would bother to access more 'accurate' information about their requirements that does take into account their personal circumstances. Even fitness apps and wearable activity monitors that give the impression of personalisation do not uniformly ask for current physical activity level and so can vastly underestimate requirements by presenting only the RMR. Calorie counting is also a slippery slope to an eating disorder in people who are vulnerable. Certainly for people in recovery from eating disorders, having calorie content information and % intakes everywhere can be difficult to escape from and hinders recovery.

Weight reduction in the 1920s

Modern FS, Johnson GE. Dietetic considerations in the treatment of obesity. California and Western Medicine. 1928;28(5):660. https://www.ncbi.nlm.nih.gov/pmc/articles/PMC1655811/pdf/calwestmed00197-0067.pdf

This paper is an interesting insight to the medical thinking at the time as it has the research paper which advocates salt and fluid restriction for weight loss (!) but also has some letters from MDs in response to the paper. It seems that charlatans looking to make a buck from weight loss products have always aggravated the medical professions. These days though a number of their own seem to have jumped on the bandwagon!

> Until comparatively recently obesity has received scant attention by the medical profession. With the increasing knowledge of its importance as a precursor of diabetes, cardiorenal-vascular diseases and the arthropathies, strenuous efforts are now being made to salvage this condition from the hands of the quacks and faddists who have long recognized it as an important source of revenue.

Special mention for the inclusion of mineral oil in their low calorie diets because the constipation was so severe!

Weight loss, weight regain pattern

Studies which follow participants for 2-5 years find that the vast majority have regained all that they had lost in the intervention.

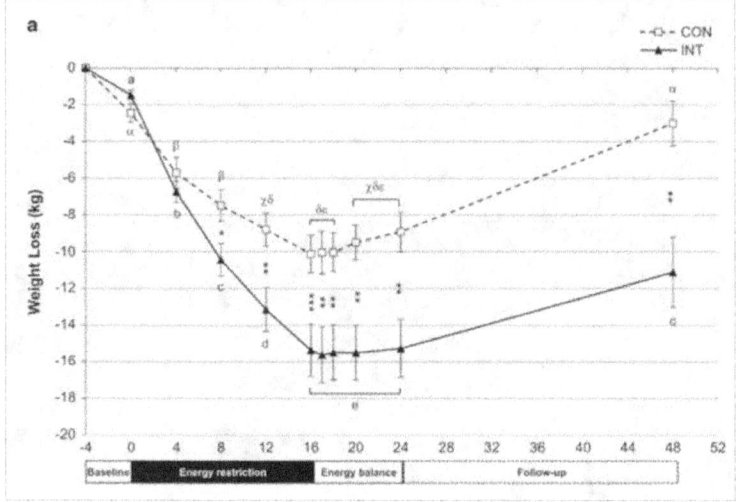

Here is what every weight loss trial looks like – weight loss until the end of the intervention period, followed by weight regain.

This one is hot off the press, from here: Byrne, Nuala M., et al. "Intermittent energy restriction improves weight loss efficiency in obese men: the MATADOR study." International Journal of Obesity 42.2 (2018): 129. https://www.nature.com/articles/ijo2017206

Episode 9: MORALITY VS ETHICS: WHY FAT IS FRAUGHT

This episode delves into theories of human morality and medical ethics and their role in fat-phobia. I unpick the concepts and meaning of fatness through these two lenses in order for listeners to be able to identify which meaning system is at play in those who force their weight-centrism upon others, and select a line of discussion that hits its mark.

Learning Outcomes:

- Recognise the key components of medical ethics – autonomy, beneficence, non-maleficence and justice
- Recognise the five foundations of morality as defined by Haidt and Graham.
- Appreciate the way that fatness, fat medicine, obesity rhetoric and cultural beliefs are influenced by ethical and moral principles dependent on perspectives and framing

Key Reading:

Principles of medical ethics – brief explainer
Gillon, Raanan. "Medical ethics: four principles plus attention to scope." BMJ: British Medical Journal 309.6948 (1994): 184.
https://www.ncbi.nlm.nih.gov/pmc/articles/PMC2540719/pdf/bmj00449-0050.pdf

Foundations of Morality Theory
Koleva, Spassena P., et al. "Tracing the threads: How five moral concerns (especially Purity) help explain culture war attitudes." Journal of Research in Personality 46.2 (2012): 184-194.
https://s3.amazonaws.com/academia.edu.documents/42957242/Tracing_the_threads_How_five_moral_conce20160222-25873-1xf0hsc.pdf?AWSAccessKeyId=AKIAIWOWYYGZ2Y53UL3A&Expires=1529904612&Signature=MkPUQrOgbJgmD%2F1MCgiaBZwxtrY%3D&response-content-disposition=inline%3B%20filename%3DTracing_the_threads_How_five_moral_conce.pdf

Moral framing of larger bodies
Friedman, May. "Fat is a Social Work Issue: Fat bodies, moral regulation, and the history of social work." Intersectionalities: A Global Journal of Social Work Analysis, Research, Polity, and Practice 1 (2012): 53-69.
http://journals.library.mun.ca/ojs/index.php/IJ/article/viewFile/350/223

Fat Medicine vs Obesity Medicine

The term 'obesity' is directly related to the arbitrary cut off at a BMI of 30, which serves no health promotion purpose. Those who identify as working in 'obesity medicine', 'obesity research' or 'obesity activism' prescribe almost entirely to an anti-obesity perspective – they are actively working to try to make larger bodies smaller, prevent bodies from getting larger and seeing larger bodies as inherently problematic. 'Fat medicine', on the other hand, refers to the provision of healthcare to people in larger bodies which aims to enhance health outcomes and treat disease effectively in the body that the person has right now, without the pursuit of weight change.

Ethical Principles in Medicine

(definitions here from Stanford University)

Autonomy
- Requires that the patient have autonomy of thought, intention, and action when making decisions regarding health care procedures. Therefore, the decision-making process must be free of coercion or coaxing. In order for a patient to make a fully informed decision, she/he must understand all risks and benefits of the procedure and the likelihood of success.

Beneficence
- Requires that the procedure be provided with the intent of doing good for the patient involved. Demands that health care providers develop and maintain skills and knowledge, continually update training, consider individual circumstances of all patients, and strive for net benefit.

Non-Maleficence
- Requires that a procedure (or intervention) does not harm the patient involved or others in society.

Justice
- The idea that the burdens and benefits of new or experimental treatments must be distributed equally among all groups in society. Requires that procedures uphold the spirit of existing laws and are fair to all players involved. The health care provider must consider four main areas when evaluating justice: fair distribution of scarce resources, competing needs, rights and obligations, and potential conflicts with established legislation. [Some] technologies create ethical dilemmas because treatment is not equally available to all people.

Here's a book chapter on these concepts (Principles of Healthcare Ethics): http://samples.jbpub.com/9781449665357/Chapter2.pdf

Moral Foundations Theory

Fairness and Reciprocity	• Basic human desire for justice and fairness • Desire for fair compensation for action (reciprocity)
Harm vs Care	• Cherishing and protecting (vulnerable) others • Protection of others from harm
In-group Loyalty	• Standing with your group, family, nation • Desire to belong, self-sacrifice for good of group
Authority and Respect	• Submitting to tradition and legitimate authority • Obedience and safety of hierarchy
Purity and Sanctity	• Religious feeling and piety, moral and physical cleanliness • Abhorrence for disgusting things, foods, actions

Moral foundation – someone's key values	Perceived challenges/risks posed by weight neutral practice messages	Solution (appeal to moral foundations to reframe)
Fairness and Reciprocity	Not acting on weight is irresponsible and unfairly neglects larger people. I work hard to stay slim and healthy and so should everyone else. Eating 'properly' will 'correct' weight. *I'm helping people access fair opportunities in life.*	Losing weight isn't the 'only' gateway to looking after yourself. It is unfair to judge someone by their appearance
Harm vs Care	Helping to people to lose weight is caring for them. *I'm helping them and they need me.*	Weight bias harms people across the lifespan. We need to do better.
In-group Loyalty	Weight loss counselling is part of what dietitians do (professional identity). *It's what feels right to me because I think like my close colleagues.*	Weight neutral professionals have fantastic collegial support networks.
Authority and Respect	I'm following the clinical guidelines when I recommend weight loss. This is what I was taught in my training. This is what I understand the research to be saying. *I'm doing what I'm supposed to do.*	Weight neutral practices are well defined, training and mentorship is available, this approach is evidence based. More programs are including it in training now, don't be left behind.
Purity and Sanctity	Assumptions about gluttony, hygiene and deviations from what's considered 'acceptable' appearance and eating behaviours. *I'm doing what's right.*	Reinforce with data that larger people do not eat 'more' or 'worse' than smaller people. Personal hygiene, dietary quality and activity patterns vary across weight categories.

Episode 10: THE NON-DIET APPROACH MODEL

In this episode I take you through the clinical model that I've developed during my PhD research. This model is a weight-neutral, size inclusive, body acceptance practice framework for clinicians.

<u>Learning Outcomes:</u>

- Appreciate the sources used to develop the Non-Diet Approach model
- Identify the 5 practice principles and theoretical framework of the Non-Diet Approach
- Identify strategies which are consistent with the non-diet approach model components

<u>Key Reading:</u>

Self Determination Theory
Deci, E. L., & Ryan, R. M. (2008). Self-determination theory: A macrotheory of human motivation, development, and health. Canadian Psychology/Psychologie canadienne, 49(3), 182-185.
https://pdfs.semanticscholar.org/a32f/3435bb06e362704551cc62c7df3ef2f16ab1.pdf

Comprehensive summary of weight-neutral approaches in dietetic practice
Marsha Hudnall MS, RDN, CD 'Mindful Eating in Nutrition Counseling', Today's Dietitian, August 2016, pages 2-9
http://viewer.zmags.com/publication/dc47a939#/dc47a939/3

Systematic review of impact of non-diet approaches
Clifford, Dawn, et al. "Impact of non-diet approaches on attitudes, behaviors, and health outcomes: A systematic review." Journal of nutrition education and behavior 47.2 (2015): 143-155.
https://www.sciencedirect.com/science/article/pii/S1499404614007969
(this is not an open access paper I'm afraid but if you contact the authors they may be able to send you a copy, otherwise the abstract does give an indication of the findings)

Comparison of diet vs non-diet approach

Traditional vs non-diet approach

www.healthnotdiets.com

Factor	Traditional	Non-Diet Approach
Self Care	Generally ignored in favor of focus on weight loss. Conditional acceptance of the self based on achievement of weight loss goals	Major focus on developing self-compassion, which enhances motivation for self care
Weight	Aim for a specific weight or % weight loss	Accept that body weight may or may not change, up or down, now or in the future.
Food	Rules from external sources, 'allowed' and 'not allowed' foods	Reliance on internal cues, body trust, mindful eating. All foods allowed
Exercise	Prescriptive exercise with the aim of weight loss	Moving for joy and meaning, no focus on weight loss
Body Image	Implies that body is unacceptable at current weight	Encourage acceptance and body respect regardless of weight

Non-Diet Approach Model

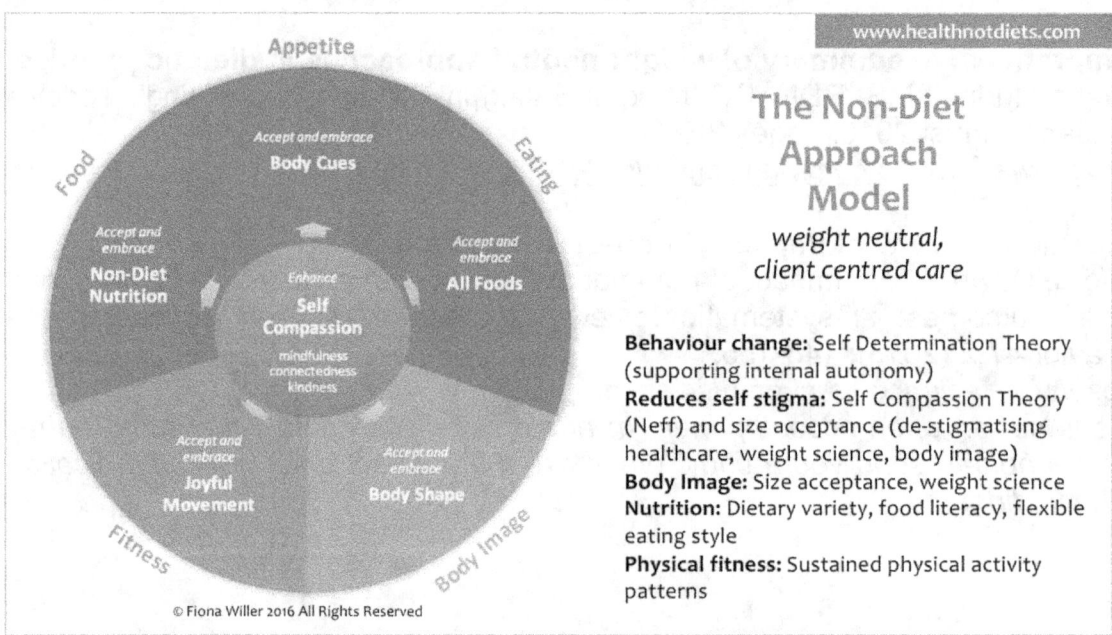

© Fiona Willer 2016 All Rights Reserved

This is available in an A4 PDF from www.healthnotdiets.com

The Non-Diet Approach guidebooks include explanation of each of the practice principles, baseline and outcome assessment tools, strategies to use with clients and worksheets to support the strategies.

Key comparisons – nourishment

Nourishment

Diet	Non-Diet Approach
Variety stipulated in meal plan or national guidelines	Variety driven by curiosity and experimentation
Guidelines translated into 'diet' Observational research reverse engineered into dietary prescription Rigid meal plan 'forcing' variety	Purpose is to enjoy the variety and experience of eating Experiment with findings of nutrition research (increase variety in oils for a few months to see if chol changes etc) Encourage enjoyment & mindfulness element Mindful eating driving variety seeking Self care through meal planning (plan for variety) Non-coercive, client decides if and when
Guilt if meal plan not followed as planned	*Disappointment if meals aren't satisfying, tasty and varied*

Key comparisons – physical activity

Physical Activity

Diet	Non-Diet Approach
Physical Activity/Exercise	Movement
Purpose is to 'burn' fat/kJ Often prescribed to the minute or calorie Used to make 'bad' food choices 'disappear' Coercive- encouraged to exercise ('should')	Purpose is to enjoy using the body in an active way Encourage enjoyment element Mindful physical activity Non-coercive, client decides if and when
Guilt if exercise not completed as planned	*Disappointment if missed*

Comparison with similar constructs/movements/programs

Weight-neutral approach comparison

Component	Health at Every Size®	Intuitive Eating™	Generic weight neutral 'lifestyle'	Non-Diet Approach	Fat activism (originated in 1960's)
Social justice/activism – working to end size discrimination and oppression in broader community and systems	✓				✓
Does not aim to control or change body weight	✓	✓	May claim 'obesity prevention'	✓	✓
Nourishment (science-informed, self-directed)	✓	✓	✓	✓	
Physical activity (science-informed, self-directed)	✓	✓	✓	✓	
Internal autonomy via principles of self determination theory	✓	✓	Internal or external	✓	✓
Body image, body dissatisfaction, internalised weight bias, own size acceptance (the individual)	✓	✓	✓	✓	✓
Self Compassion building (resilience, positive self-regard)	✓			✓	
Counselling styles: Motivational Interviewing, ACT			MI/ACT/CBT	✓	
Informed by scientific research	✓	✓	✓	✓	Ethics/ justice/law/p hilosophy

Items in pale text indicate probable weight centrism – check program carefully

Measuring outcomes (always important!)

Measuring the non-diet approach

Non-diet Approach element	Academic construct
Self Compassion Experiential Learning Mindfulness	Self Compassion Theory Self-determination Theory Mindfulness
Accept and Embrace Body Cues	Dietary Restraint
Accept and Embrace All Foods	Dietary Quality & Variety
Accept and Embrace Body Shape	Body Dissatisfaction Weight Control Beliefs
Accept and Embrace Movement	Physical activity level Enjoyment of physical activity
Accept and Embrace Non-Diet Nutrition	Dietary Quality Enjoyment of food and eating

Resources - books

Non Diet Approach Guidebook for Dietitians Fiona Willer, AdvAPD

Intuitive Eating Evelyn Tribole & Elyse Resch, RDs

If not dieting, then what? Dr Rick Kausman

Health at Every Size: The Surprising Truth About Your Weight Linda Bacon, PhD

Eat What You Love, Love What You Eat (series) Dr Michelle May & others

Body of Truth Dr Harriet Brown

Body Respect Linda Bacon

Diet Survivors Handbook Judith Matz

Nourish Heidi Schauster

Episode 11: UNBOXING THE SCIENCE OF WELLNESS MARKETING

As my following on Instagram grew I started getting approached by 'health and wellness' brands to spruik their products, and one lot even sent me some samples of their product to try – without asking if I wanted them! They came with a lovely letter and some product information, filled to the brim with reasons why their product should be in every home across the country. So I'm going to dissect their arguments for your listening pleasure and highlight the tricks used in wellness-food marketing to persuade us to open our wallets and our mouths for just about anything.

Learning Outcomes:

- Develop an understanding of the legal limitations on marketing 'wellness' foods
- Identify key messages used by wellness product marketers to persuade the public to buy their product
- Recognise the techniques used in wellness marketing.

Key Reading:

Apple cider vinegar health effects
https://theconversation.com/is-apple-cider-vinegar-really-a-wonder-food-86551

Confessions of an Instagram Influencer
Eye-opening! Let the record show that this is NOT how I built my (relatively small) Instagram following! Obviously it's not a scientific research paper but still good info to know.
https://www.entrepreneur.com/article/315156

Social media marketing research
Erdoğmuş, İrem Eren, and Mesut Cicek. "The impact of social media marketing on brand loyalty." Procedia-Social and Behavioral Sciences 58 (2012): 1353-1360.
https://s3.amazonaws.com/academia.edu.documents/50672704/The_Impact_of_Social_Media_Marketing_on_20161202-26122-48gyp8.pdf?AWSAccessKeyId=AKIAIWOWYYGZ2Y53UL3A&Expires=1533358711&Signature=F1GuDumKLIZ07mWYUDq%2BQAH2kmw%3D&response-content-disposition=inline%3B%20filename%3DThe_Impact_of_Social_Media_Marketing_on.pdf

Food Labelling – Food Standards Code (Australia)

Health claims refer to a relationship between a food and health rather than a statement of content. There are two types of health claims:

- **General level health claims** refer to a nutrient or substance in a food, or the food itself, and its effect on health. For example: **calcium for healthy bones and teeth**. They must not refer to a serious disease or to a biomarker of a serious disease.
- **High level health claims** refer to a nutrient or substance in a food and its relationship to a serious disease or to a biomarker of a serious disease. For example: **Diets high in calcium may reduce the risk of osteoporosis in people 65 years and over**. An example of a biomarker health claim is: **Phytosterols may reduce blood cholesterol**.

http://www.foodstandards.gov.au/consumer/labelling/nutrition/Pages/default.aspx

Advertising and sales – Australian competition and consumer commission (ACCC)

Businesses are not allowed to make statements that are incorrect or likely to create a false impression.

This rule applies to their advertising, their product packaging, and any information provided to you by their staff or online shopping services. It also applies to any statements made by businesses in the media or online, such as testimonials on their websites or social media pages.

For example, businesses cannot make false claims about:

- the quality, style, model or history of a product or service
- whether the goods are new
- the sponsorship, performance characteristics, accessories, benefits or use of products and services
- the availability of repair facilities or spare parts
- the need for the goods or services
- any exclusions on the goods and services.

It makes no difference whether the business intended to mislead you or not. If the overall impression left by a business's advertisement, promotion, quotation, statement or other representation creates a misleading impression in your mind—such as to the price, value or the quality of any goods and services —then the behaviour is likely to breach the law.

There is one exception to this rule. Sometimes businesses may use wildly exaggerated or vague claims about a product or service that no one could possibly treat seriously or find misleading. For example, a restaurant claims they have the 'best steaks on earth'. These types of claims are known as 'puffery' and are not considered misleading.

https://www.accc.gov.au/consumers/misleading-claims-advertising

For products that are claimed to be 'therapeutic' but are NOT food, the Therapeutic Goods Act is relevant too. Our Apple Cider Vinegar is considered food so does not come under this law, but Apple Cider Vinegar tablets (which do exist!) would. https://www.tga.gov.au/who-we-are-what-we-do

Functional food

Here is a decent plain language summary of functional foods, remembering that it is a completely arbitrary label because the more we study a food the more likely it is that we discover something other than basic nutrients in it, and producers are keen to add things to their products to be able to call them functional foods (vitamin water, anyone?):

> **Currently there is a lot of buzz about functional foods, but what exactly are functional foods?**
>
> According to the Food and Drug Administration (FDA), there is no official definition for functional foods. The Academy of Nutrition and Dietetics define functional foods as "whole foods along with fortified, enriched or enhanced foods that have a potentially beneficial effect on health when consumed as part of a varied diet on regular basis at effective levels based on significant standards of evidence."
>
> Basically, functional foods are foods that provide a health benefit in addition to macro and micronutrients. These foods are vital in disease prevention and include fortified foods, phytonutrient-containing fruits and vegetables, fermented foods, fish and chocolate.

See the whole article here: https://www.forbes.com/sites/quora/2018/05/22/what-are-functional-foods/#78b707e33dc4
Also good: https://www.eatright.org/food/nutrition/healthy-eating/functional-foods

Push and Pull Marketing

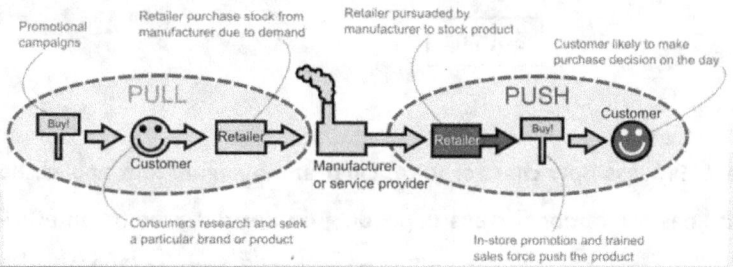

Push or Pull Strategies

Push Strategies producers use sales promotions to push products through the food distribution channel
- Uses trade-oriented promotions
- Negotiation with retailers to stock your product
- Point of sale displays

Pull Strategies aim to induce *consumers* to purchase and request more
- the information of increased (or initial) demand works its way upstream towards the producer
- consumers *pull* product towards themselves
- Advertising and mass media promotion
- Sales promotions and discounts ; Word of mouth referrals; customer relationship management.

http://sites.psu.edu/agbm302/wp-content/uploads/sites/37775/2016/04/Promotion-Strategy.pdf

How is apple cider vinegar made?

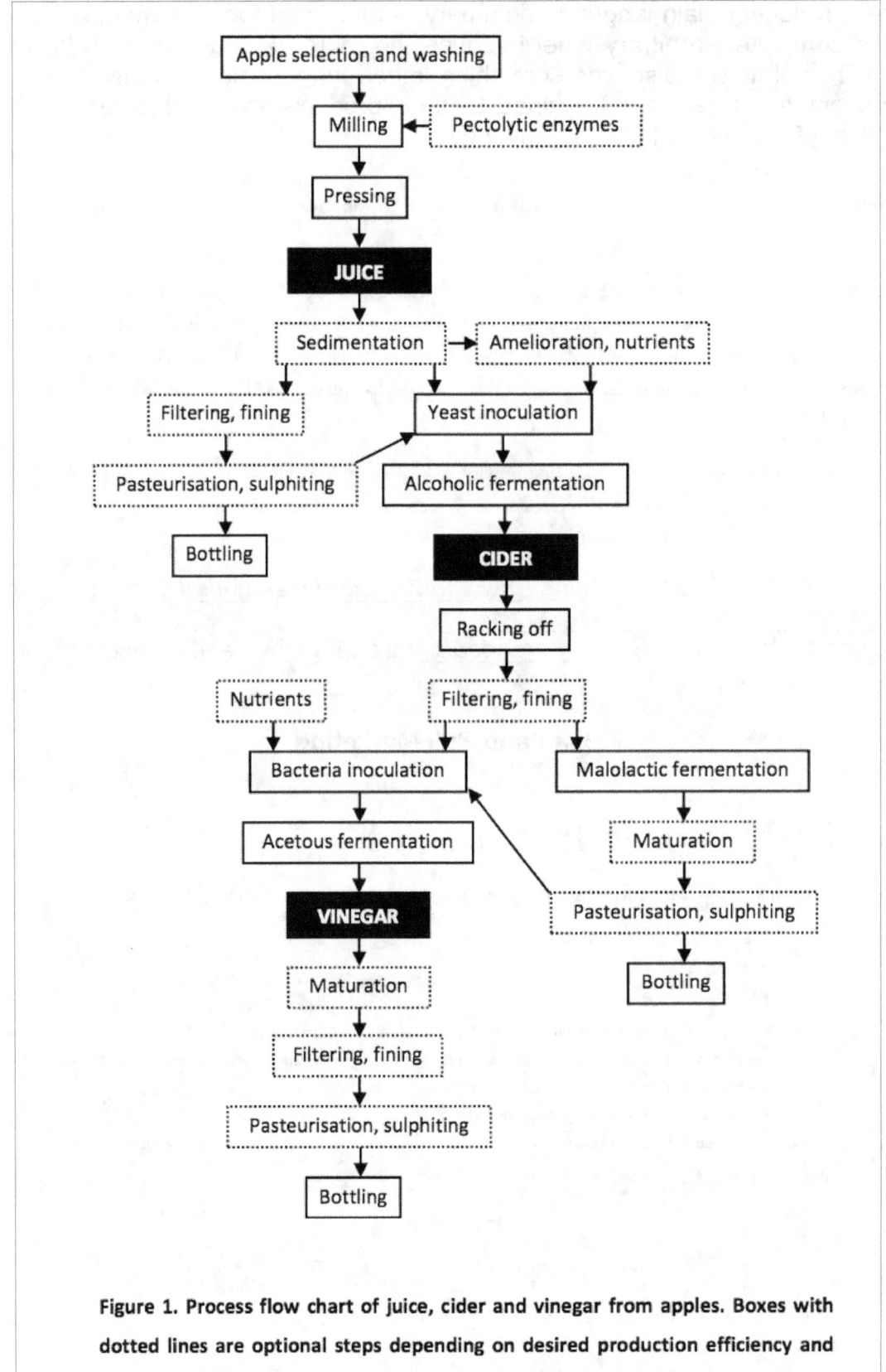

Figure 1. Process flow chart of juice, cider and vinegar from apples. Boxes with dotted lines are optional steps depending on desired production efficiency and quality characteristics. Chart modified from Joshi & Sharma (2009).

From: https://stud.epsilon.slu.se/2481/1/heikefelt_c_110415.pdf

Complementary and Alternative Medicine (CAM)

CAM therapy classification	Examples
Holistic medical systems	*Homeopathic medicine, traditional medicine, Ayurveda*
Mind-body medicine	*Patient support groups, cognitive-behavioral therapy, prayer, mental healing, creative outlet therapies*
Biologically based practices	*Herbs, food, vitamins, dietary supplements, herbals, use of shark cartilage to treat cancer*
Manipulative + body-based practices	*Chiropractic medicine, massage, osteopathic manipulation*
Energy medicine	*Bio-field therapies, qi gong, Reiki, therapeutic touch, bio-electromagnetic based treatment*

CAM users and top conditions that CAM is used with

Traits associated with CAM use

Female[11, 12, 13, 14]

Middle aged[13, 14]

Higher levels of spirituality[15, 16]

Lower emotional role functioning[11]

Lower perceived health[11, 15]

Serious, chronic, or longer illness[14]

Higher level of education[11, 13, 15]

Previous transformational experience leading to a worldview change[15]

Holistic view of health problems[15]

Top medical conditions treated with CAM[13]

Back pain and related pathology

Depression

Insomnia

Headache or migraine

Stomach or intestinal illness

From:
Rebecca M. Widder, Douglas C. Anderson, The appeal of medical quackery: A rhetorical analysis, Research in Social and Administrative Pharmacy, Volume 11, Issue 2, 2015, Pages 288-296, ISSN 1551-7411,
https://doi.org/10.1016/j.sapharm.2014.08.001.
(http://www.sciencedirect.com/science/article/pii/S1551741114002873)

Episode 12: WHAT-THE-FIVEPERCENT?? TYPE 2 DIABETES & WEIGHT LOSS RECOMMENDATIONS

We continually hear this magical 5% weight loss goal for people with type 2 diabetes, that losing 5% of your body weight is going to result in protecting you from all sorts of terrible diabetes related things…. In this episode I talk about where this figure has come from, why it misconstrues its own science, and weight neutral ways to manage diabetes and the diabetes healthcare environment.

Learning Outcomes:

- Develop an understanding of the life course of type 2 diabetes
- Identify the reasons that 5% weight loss is considered a good thing for people with type 2 diabetes
- Recognise the flaws in the research assumptions and methods relating to type 2 diabetes and weight loss.

Key Reading:

Type 2 diabetes life expectancy tables
Leal, Jose, Alastair M. Gray, and Philip M. Clarke. "Development of life-expectancy tables for people with type 2 diabetes." *European heart journal* 30.7 (2008): 834-839.
https://www.ncbi.nlm.nih.gov/pmc/articles/PMC2663724/

Weight loss recommendations – for critique
Wilding JPH. The importance of weight management in type 2 diabetes mellitus. International Journal of Clinical Practice. 2014;68(6):682-691. doi:10.1111/ijcp.12384.
https://www.ncbi.nlm.nih.gov/pmc/articles/PMC4238418/

Lifestyle management – weight neutral
Weisenberger, Jill. 'Diabetes Management & Nutrition Guide: Lifestyle Strategies for Reversing Prediabetes' *Today's Dietitian* (2018) Vol. 20, No. 8, P. 48
http://www.todaysdietitian.com/newarchives/0818p48.shtml

Development and progression of Type 2 Diabetes

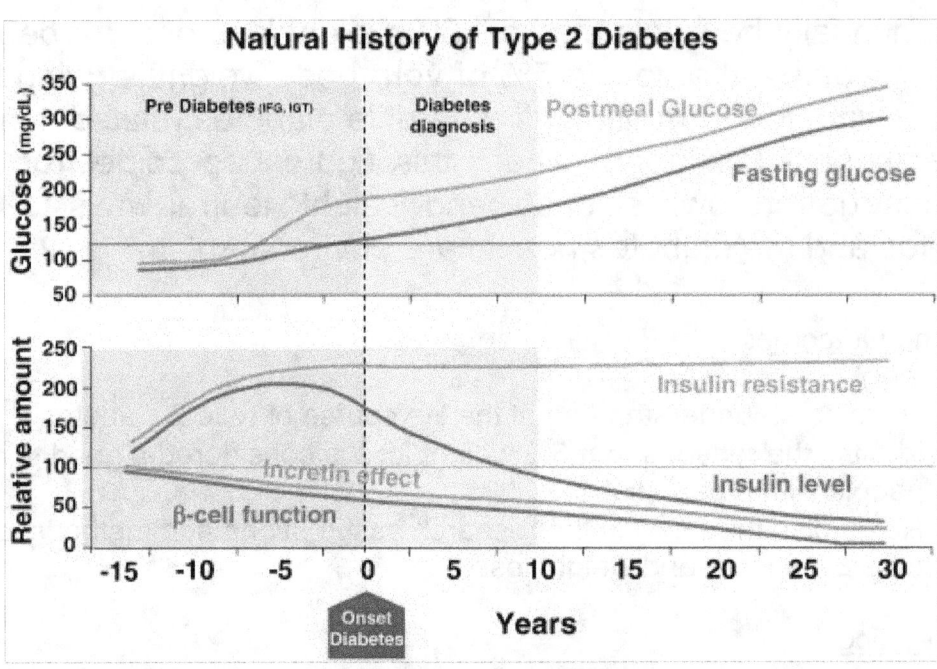

http://diabetes.diabetesjournals.org/content/58/4/773

Comparison of prevalence in BMI categories between common chronic conditions

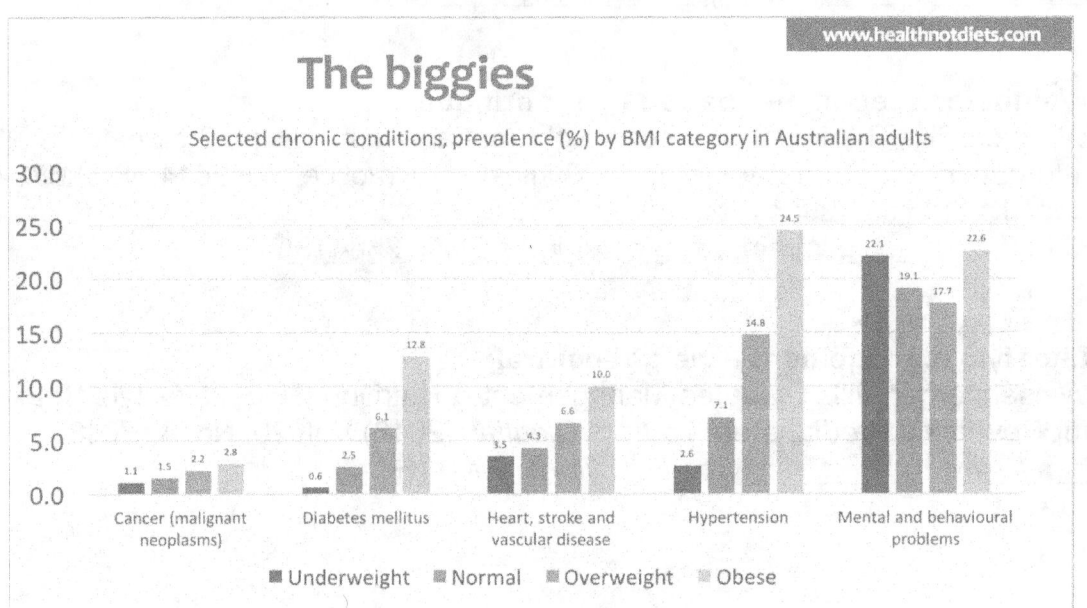

Australian Bureau of Statistics release 4364055001DO005_20142015 National Health Survey: First Results, 2014–15

Type 2 diabetes recommendations: Clinical Guidelines for management of Overweight and Obesity in Adults, Adolescents and Children (2013)

Table C9 Weight change and glycaemic control in adults with type 2 diabetes following lifestyle intervention

LEVEL I STUDIES

Study	Intervention	Weight change	HbA1c
Norris et al. 2005b* $n = 4659$; 5 years of follow-up	Lifestyle interventions Diet vs usual care	Up to ↓12.0 kg ↓3.0 kg	↓0.7%
Nield et al. 2007 $n = 1467$; 12 months of follow-up	Diet and exercise	↓2.5–5 kg	↓1.0%
Huisman et al. 2009 $n = 5469$	Lifestyle interventions at < 6 months Lifestyle interventions at > 6 months	Effect size 0.18 Effect size 0.06	Effect size 0.35 Effect size 0.34
Thomas et al. 2006 $n = 377$ 12 months of follow-up	Exercise vs no exercise	0.0 kg	↓0.6%

Cochrane review, should be listed as 2 year follow up

LEVEL II STUDIES

Study	Intervention	Weight change	HbA1c	Blood pressure (mmHg)		Lipids (mg/dL)		
				Systolic	Diastolic	LDL	HDL	Tg
Belacazar et al. 2010 Mean age: 57.5 $n = 1759$ 5 years of follow-up	Intensive lifestyle intervention	↓9.0 kg	↓0.7%	—	—	↓0.07	↑0.2	↓1.8
	Diabetes self-education	↓0.8 kg	—	—	—	↓0.3	↑0.08	↓0.7
Pi Sunyer et al. 2007 Age: 45–74 $n = 5145$ 12 months of follow-up	Intensive lifestyle intervention	↓8.6%	↓1.3%	↓6.8	↓3.0	↓0.3	↑0.2	↓1.7
	Diabetes self-education	↓0.7%	↓0.1%	↓2.8	↓1.8	↓0.3	↑0.08	↓0.8
Wing 2010a Mean age: 57.5 $n = 5145$ 4 years of follow-up	Intensive lifestyle intervention	↓6.5%	↓0.36%	↓5.33%	↓2.92%	↓0.6	↑0.2	↓1.4
	Diabetes self-education	↓0.88%	↑0.09%	↓2.97%	↓2.48%	↓0.7	↓0.1	↓1.1
Cheskin et al. 2008 Mean BMI: 35 $n = 119$ 86 weeks of follow-up	Portion controlled diet	↓5.6 kg	No change	↓7.6%	↓2.7%	—	↑0.2	↑0.02
	Standard diet	↓4.7 kg	↓1.2%	↓14.0%	↓9.7%	—	↑0.4	↓0.15

This one reports the outcomes at 12 months, not 5 years

These two are from the same study

LEVEL II STUDIES

Study	Intervention	Weight change	HbA1c	Blood pressure (mmHg)		Lipids (mg/dL)		
				Systolic	Diastolic	LDL	HDL	Tg
Christian et al. 2008** Mean BMI: 35.4 $n = 310$ 12 months of follow-up	Nutrition and physical activity counselling	↓0.2 kg	↓0.14%	↓2.6%	↓2.6%	—	↓0.02	↓0.8
	Health education pamphlet	↑1.4 kg	↓0.46%	↓4.7%	↓2.5%	—	↑0.09	↓0.5

— = not measured; HbA1c = glycated haemoglobin; HDL = high-density lipoprotein; LDL = low-density lipoprotein; mg/dL = milligrams per decilitre; mmHg = millimetres of mercury; Tg = triglycerides

* More intense physical activity was associated with greater weight reduction but not with greater reductions in HbA1c.

** Because 98% of participants were taking antihyperglycaemic medications, the effect of medication use on HbA1c was unable to be controlled for.

https://www.nhmrc.gov.au/guidelines-publications/n57 p148

How did we get so hung up on the 5% weight loss goal?

In diabetes weight loss trials, it was observed that people who lost 5% or more weight also tended to have improvements to their blood sugar levels and other markers of metabolic health such as cholesterol levels and blood pressure.

People in the trials who didn't lose weight were assumed to be non-compliant (as in, they didn't do the things that were asked of them) rather than thinking of them as non-responders (as in, they were doing the things, but their body/metabolic circumstances meant that the things didn't result in weight loss

---

for them). Trials get themselves in a paradox loop when they use their primary outcome measure also as a measure of fidelity to the program.

Remember the effects of energy restriction (aka starvation, aka 'metabolic austerity') in a human (with significant variations between individuals):
- Blood pressure reduces
- Insulin sensitivity increases
- Blood glucose levels decrease
- Weight loss
- Preoccupation with food and eating increase to the detriment of other interests
- Gastrointestinal discomfort
- Decreased sex drive
- Skeletal and organ muscle depletion/wasting
- Declines in bone density
- Stunted growth if foetus/child/adolescent

Remember that with any weight loss intervention, the things that are doing the work but which are attributed to weight loss, are:
- Energy restriction (as above)
- Dietary changes
- Physical activity changes
- Receiving care and concern from research team and/or participant support group, which lowers self-stigma and stress

Reading the research

Within group analysis – this means they looked at what happened to the people in the intervention group separately from the beginning to the end of the study – this type of analysis can look at whether HbA1c and cholesterol level changes were related to weight loss. The people that end up losing weight and have concurrent changes in their biochemistry can be thought of as 'weight loss responders'. These kinds of studies usually attribute failure to lose weight as an adherence or compliance problem – it is assumed that they didn't do the things asked of them – rather than acknowledge that their participants have complicated histories and metabolic circumstances – and we're not really great at measuring, or controlling for everything, yet.

Between group analysis – this means that the trial had an intervention group (that got the 'active program') and a control group (that might have just done nothing, or have received 'standard treatment'). To see if the intervention had a 'real' effect, rather than coinciding with effects that happen due to time passing, the participants are randomised into either the intervention group or the control group at the start (so they've got two theoretically identical groups) and then their results are compared at the end – so they look at the outcomes in each group (the final measures taken) and compare them, they can then say that the intervention was or wasn't effective at getting people to lose weight or improve their blood sugars or whatever they're interested in. In this analysis design it means that the associations between measures in each group are divorced though – so someone with a benefit to their cholesterol but with no weight loss, will be counted towards the 'improvements' in the weight loss group – so it looks like weight loss was the thing, but it was some other element for that person.

Weight loss is usually used as a proxy (stand in/surrogate marker) for energy restriction because it's easier to measure, but the problem is that it's easy to start believing that it's the weight loss and not the energy restriction that was the 'active ingredient'. This means that we get the impression that if weight losses could be maintained, then the other benefits should be maintained as well, except that when we remember that it is the energy restriction doing the work we realise that staying in a state of energy restriction indefinitely is not reasonable or feasible.

So what can you do when you're given the recommendation to lose weight?

- 'I'm happy to tweak my lifestyle but I'm not going to starve myself'
- 'What treatment would you offer someone in my situation if my BMI was 22?' Would they get offered medication?
- 'Could you please write me a referral for a HAES/non-diet dietitian or diabetes educator?'
- Tell them you've already lost 5% of your body weight and ask, what's next?
- 'Having a weight loss intention doesn't help me look after myself – I'm going to focus on learning how my body responds to food and physical activity' – is there anything else you can offer me?'

Other things to ask your primary care provider:
- 'How frequently do I need to come back for check-ups?'
- 'What symptoms do you want me to come in especially for?'
- 'Can you point me towards any local or online peer support groups that aren't all about dieting?' (not because they're going to know any – it's to gently indicate that they SHOULD know about some)

In my experience, telling a healthcare professional about your history of an eating disorder does not automatically result in them realising that prescribing weight loss is inappropriate and dangerous – they should realise, but dieting is so normative that unless they have engaged in this stuff they don't typically get it. If you're a health, fitness or counselling professional who is communicating with your clients primary care practitioner, you're going to need to spell this out 'Due to XX's eating disorder history, weight loss goals are contraindicated/not appropriate at this time.'

Episode 1: Stuck in a weight-centric operating system

1. Weight centric perspectives are currently prioritised in which domains?
   a. 'Obesity' research
   b. Public Health Campaigns
   c. Clinical primary care practice
   d. All of the above

2. Reading a story about someone's weight loss journey and feeling vindicated about your own positive attitude towards people in smaller bodies is an example of
   a. Cognitive Bias
   b. The Dunning-Kruger Effect
   c. Confirmation Bias
   d. Selective Recall

3. Professional competence is enhanced when
   a. Clinicians have an accurate working knowledge of all treatment options
   b. Clinicians have examined their own cognitive biases and worked through how issues relevant to their own personal experience may influence the therapeutic relationships with their clients
   c. They follow clinical guidelines with no consideration of the potential psychological impact on their clients
   d. a. and b.

4. Cognitive Dissonance refers to:
   a. a systematic pattern of deviation from norm or rationality in judgment. Individuals create their own "subjective social reality" from their perception of the input.
   b. the mental discomfort (psychological stress) experienced by a person who simultaneously holds two or more contradictory beliefs, ideas, or values.
   c. a motivational reaction to offers, persons, rules, or regulations that threaten or eliminate specific behavioural freedoms.
   d. extreme anxiety, sorrow, or pain.

5. According to Bombak (2014), alternatives to the seemingly flawed weight-focused approach include:
   a. Health at Every Size (HAES)
   b. Health in Every Respect (HIER)
   c. Healthy at Any Size (HAAS)
   d. a. and b.

Episode 2: Demystifying Definitions, De-Myth-Defying Assumptions

1. Weight neutral synonyms include:
   a. Weight inclusive, size accepting
   b. Health promotion, weight management
   c. Health at Every Size, HAES
   d. a and c

2. A 'lifestyle' intervention always includes advice, content and or activities regarding:
   a. Food or eating
   b. Physical activity or body movement
   c. Psychological coping and strategies to support behaviour change
   d. All of the above

3. Which of the following factors is present in all weight neutral intervention models?
   a. Supports personal autonomy
   b. Size acceptance
   c. Nourishment through variety
   d. Timely, appropriate medical care

4. 'Weight neutral' does NOT mean that:
   a. elements proven superior in the care of particular body sizes are ignored
   b. body weight is invisible or irrelevant
   c. the weight-related experiences of people are erased
   d. weight-related bias, stigma, discrimination and oppression are denied
   e. all of the above.

5. According to Tylka et al (2014), problematic individual factors associated with a weight-normative or weight centric approach for people with larger bodies are:
   a. Weight cycling, binge eating, depression
   b. Binge eating, obesity, oppression
   c. Poverty, depression, learned helplessness
   d. Negative weight bias, larger is considered less healthy.

Episode 3: How we got here: BMI meets death

1. The Quetelet Index which we now refer to as Body Mass Index, was devised in:
   a. 1869
   b. 1959
   c. 1996
   d. 1851

2. All-cause mortality risk by Body Mass Index uses the following information:
   a. BMI, age at death, lifestyle factors like dietary intake and physical activity
   b. BMI, age, sex, age at death
   c. Height, one adult weight, age at death
   d. Height, weight variation across adult life, sex, age at death

3. An American adult with a BMI of 27, depending on the year, would have been variously classified as:
   a. 'Ideal weight', 'overweight', 'obese'
   b. 'desirable weight', 'average weight', 'overweight'
   c. '50th percentile', 'median weight', 'at risk of overweight'
   d. 'Normal weight', 'overweight', 'pre-obese'

4. During adolescence, growth in height and weight of an individual is characterised by:
   a. A growth spurt of height and weight around the 14th year
   b. Independent height and weight spurts, each peaking within a 3 year period
   c. Stable rate of growth from the 11th to 17th year
   d. The pattern as depicted on the 50th centile of the CDC and WHO growth charts

5. The World Health Organisation childhood growth charts depict:
   a. The growth of all kids from one particular population as they grew over time
   b. The averages of all the studies of kids' growth that have ever been
   c. Growth of 'healthy', food-secure kids in community studies in 6 countries
   d. That all children should aspire to be on the 50th centile as they grow

Episode 4: Weight Bias, Stigma and Discrimination

1. Select the example/s of weight discrimination:
   a. Defined or default healthcare pathways where body size determines access to services
   b. Receiving verbal abuse about your body size
   c. Negative portrayal of larger bodies by the media, for example with heads out of shot
   d. a) and c)

2. Weight bias is:
   a. The negative social impact of weight discrimination
   b. A preference for or against particular body weights or shapes
   c. Persistent devaluing of the self and self-censorship of life activities based on higher body weight or shape
   d. Against the law in many westernised countries

3. Which of the following occurs as a result of weight bias:
   a. Missed identification of illness due to symptoms being dismissed due to weight
   b. Overtreatment of people with higher weights via repeated referrals to weight loss treatments and services
   c. Weaker evidence base for medication dosage and effectiveness for larger bodies
   d. All of the above

4. In the study listed by Puhl and colleagues, the top ranked strategy selected by women for addressing weight stigma in healthcare and medical settings was:
   a. Medical school education should be required to teach students about weight stigma
   b. Healthcare providers should receive training to provide more respectful, compassionate care to patients with obesity
   c. Medical schools should be required to provide a comprehensive education about obesity to medical students
   d. Healthcare providers should be educated about weight stigma and its harmful impact on people who have obesity

5. In the study listed by Sutin and colleagues, the experience of weight discrimination was associated with:
   a. Moving from a BMI of <30 to a BMI of > 30 during the study period
   b. Moving from a BMI of <25 to a BMI of >30 during the study period
   c. Remaining in the >30 BMI category during the study
   d. a) and c)

Episode 5: The Anatomy of a Weight Loss Paper

1. Which section is the least scientifically reliable in a weight loss research paper:
   a. Introduction
   b. Methods
   c. Results
   d. Discussion
   e. Conclusion

2. Weight loss intervention studies could be more accurately described as:
   a. Starvation studies
   b. Lifestyle studies
   c. Studies of combined dietary, physical activity and cognitive strategies to elicit an energy deficit
   d. Training in healthy eating

3. Which of the following is not a surrogate marker:
   a. Blood sugar level
   b. Blood cholesterol levels
   c. High blood pressure
   d. 7 meter walking test
   e. Foot amputation

4. Which of the following features increases a study's reliability?
   a. An attrition (drop out) rate of more than 20%
   b. Using an Intention to Treat (ITT) analysis method
   c. Using a within group analysis for an intervention group
   d. Using a last measurement carried forward analysis method
   e. Having a one month follow up period

5. In the paper listed by Aphramor, findings included:
   a. Dietetic literature on weight management fails to meet the standards of evidence based medicine
   b. There is a lack of academic debate on the ethical implications of continuing to promote ineffective treatment regimes
   c. The dominant view in current obesity discourse asserts that weight loss per se is synonymous with improved health
   d. All of the above

Episode 6: The Science of Self-Compassion

1. Self-compassion as defined by Neff consists of which three constructs?
    a. Mindfulness-Self Kindness, Isolation-Common Humanity, Overidentification-self judgement
    b. Isolation-Mindfulness, Self Kindness-Self Judgement, Common Humanity-Overidentification
    c. Self Judgement-Self Kindness, Isolation-Common Humanity, Overidentification-Mindfulness
    d. Common Humanity-Isolation, Self Kindness-Mindfulness, Self Jugdement, Overidentification

2. Self-compassion is useful for dealing with:
    a. Emotional, physical and circumstantial pain and suffering
    b. Joy and happiness
    c. Existential angst
    d. a) and c)

3. Symptoms of compassion fatigue include:
    a. Anxiety
    b. Sleep disturbance
    c. Self-doubt
    d. Withdrawal/isolation
    e. All of the above

4. High self-compassion is reflected with a Self Compassion Scale (Neff) score of:
    a. 1.0-1.5
    b. 1.5-2.5
    c. 2.5-3.5
    d. 3.5-5.0
    e. 4.0-5.0

5. The modern academic conceptualisation of self-compassion has roots in which tradition?
    a. Islam
    b. Christianity
    c. Buddhism
    d. Confucianism

Episode 7: Weight-Neutral, Health-Enhancing Habits

1. Which of the following would NOT be considered a risk factor related to **exposure** ?
    a. Sunlight and sunburns
    b. Second-hand cigarette smoke
    c. Kidney failure
    d. Weight stigma

2. Which of the following would NOT be considered a **surrogate marker** of disease/disease risk?
    a. Heart attack/myocardial infarction
    b. Blood sugar level peak
    c. Lung function test
    d. Blood cholesterol levels

3. Which of the following are valid indicators of health?
    a. Age at death
    b. Disease status
    c. How you feel in your body
    d. Physical function
    e. All of the above

4. In likely order of likely proportional impact, which of these are key health enhancing habits?
    a. Eating 5+ fruit/veg, being physically active, sleeping more than 8 hours/day
    b. Not smoking, sleeping 5-8 hours/day, being physically active
    c. Feeling connected with your community, eating 5+ fruit/veg, not smoking
    d. Not smoking, being physically active, feeling connected with community

5. In the meta-analysis by Aune et al (2017) listed above, how many studies were included?
    a. 95
    b. 142
    c. 600
    d. 407

Episode 8: Theoretical vs Actual Weight Change: Industrialised Wishful Thinking

1. Which of the following is NOT the name for an equation used to calculate energy requirements?
   a. Schofield
   b. Kaplan Meier
   c. Mifflin St Jeor
   d. Harris-Benedict

2. According to the NHMRC, behavioural weight losses are maximal at 6-12 months and by _____ most are back to their pre-intervention weight.
   a. 2-5 years
   b. 1-2 years
   c. 12-18 months
   d. 5-10 years

3. The most accurate method for estimating the energy needs of an individual is:
   a. Calculating how much energy from food they eat
   b. Testing using doubly labelled water
   c. Calculating their RMR using a predictive equation and multiplying by an activity or injury factor
   d. Using the average adult dietary intake figure provided on food labelling

4. In the paper by Modern and Johnson (1928) which of the following 'obesity treatments' was not favoured by the authors?
   a. Severe dietary energy restriction
   b. Sweating and restricting salt and water intake
   c. Routine thyroid medication
   d. Copious exercise

5. A mean weight loss of 5.4kgs for a group of 100 adults over 12 weeks with a standard deviation of 4kgs would indicate:
   a. Most people lost close to 5.4kgs, and this can be expected for others who try the protocol
   b. Weight losses varied significantly, with about 13 people gaining up to 2.6kgs and about 13 people losing between 9.4 and 13.4kgs.
   c. If they continued the diet until 24 weeks the group would have lost an average of 10.8kgs
   d. b and c

Episode 9: Morality Vs Ethics: Why Fat Is Fraught

1. Which moral foundation/s are being expressed in the statement 'It's unhealthy to be fat'
   a. Purity/sanctity and harm/care
   b. In-group loyalty and authority/respect
   c. Fairness/reciprocity and harm/care
   d. Purity/sanctity and in-group loyalty

2. A response to the statement in question 1 using a consistent moral foundation would be:
   a. 'All unhealthy people deserve care.'
   b. 'The new medical guidelines say that being fat isn't as unhealthy as you might think.'
   c. 'We all, including you probably, have health challenges, healthy - unhealthy isn't really a clear thing for anyone.'
   d. 'Assuming that all fat people are unhealthy isn't really fair.'

3. In the paper by Friedman, the author argues that:
   a. Fat bodies are treated identically to other oppressed bodies.
   b. Common sense logic often leads to appallingly bad policy.
   c. Fat people cannot possibly hold society responsible for the negativity they experience since they choose to be that way.
   d. Using the average adult dietary intake figure provided on food labelling

4. The type of 'intuitive ethics' that has the most impact when compared with the other categories is:
   a. Purity/Sanctity
   b. Authority/Respect
   c. In-group loyalty
   d. Harm/care
   e. Fairness/reciprocity

5. The concept of 'fat medicine' includes:
   a. Encouraging population weight control via public health messaging
   b. Encouraging and supporting health enhancing behaviours for all
   c. Provision of medical services based on clinical need and reflective of the local community's needs
   d. Research into effective screening, diagnostics and treatments for people with larger bodies, without trying to elicit weight change
   e. b, c and d

Episode 10: The Non-Diet Approach Model

1. Which of the following is not part of The Non-Diet Approach model?
   a. Accept and embrace clean living
   b. Accept and embrace body cues
   c. Accept and embrace non-diet nutrition
   d. Accept and embrace all foods

2. The underlying theoretical framework for behaviour change in the non-diet approach is:
   a. Maslow's Hierarchy of Needs
   b. The Transtheoretical Model of Change
   c. Self Determination Theory
   d. Precede-Proceed Model

3. When considering the concepts described by Deci et al, a weight loss goal is most accurately described as:
   a. An intrinsic aspiration using controlled orientation
   b. An extrinsic aspiration using impersonal orientation
   c. An intrinsic aspiration using autonomous orientation
   d. An extrinsic aspiration using controlled orientation

4. The Non-Diet Approach is:
   a. A manualised treatment specifically for larger people
   b. A clinical practice model for weight-neutral, size acceptance approaches
   c. A lifestyle program
   d. a. and c.

5. Thinking about the concepts in 'accept and embrace body shape', which of the following academic metrics would be the most useful to measure change in this domain over time in non-diet approach clients?
   a. Body Appreciation Scale-2 (BAS2)
   b. Intuitive Eating Scale-2 (IES2)
   c. Rapid Eating and Activity Assessment (REAP)
   d. International Physical Activity Questionnaire (IPAQ)

Episode 11: Unboxing the Science of Wellness Marketing

1. A dietitian posting on Instagram about a new brand of high fibre muesli bars that she tried recently and asking her followers if they've tried them too is an example of:
   a. Push marketing
   b. Shove marketing
   c. Pull marketing
   d. Passive marketing

2. Complementary and Alternative Medicine (CAM) has been found to appeal most to:
   a. Middle aged, middle class women
   b. People with chronic conditions that can be difficult to treat
   c. People who have other anti-establishment beliefs
   d. All of the above

3. The 'mother' or SCOBY used in vinegar making tends to contain predominantly what type of organism:
   a. Bacteria
   b. Yeasts
   c. Fungi
   d. Pathogens

4. Consuming vinegar, and by extension consuming apple cider vinegar has the effect of:
   a. Weight loss
   b. Heart disease protection
   c. Curing some forms of cancer
   d. Slowing the rise in blood sugar levels after a high GI carbohydrate-rich meal

5. 'Functional Foods' are those that have:
   a. Been modified to have increased amounts of a health-supporting ingredient
   b. Additional health benefits beyond basic nutrition
   c. Medicinal properties that help control particular health conditions
   d. Been developed for use in particular medical circumstances
   e. Any of the above

Episode 12: What-The-5percent?? Type 2 Diabetes & Weight Loss Recommendations

1. Which metric is of primary importance in the management of Type 2 Diabetes?
   a. Blood cholesterol
   b. Blood sugar levels
   c. Blood triglycerides
   d. Urinary glucose

2. Many of the benefits ascribed to weight loss for people with type 2 diabetes are actually the result of:
   a. Temporary overall energy restriction
   b. Improved physical fitness
   c. Consuming greater amounts of fruit and vegetables
   d. Reduced carbohydrate load while dieting
   e. All of the above

3. According to the paper by Leal et al, the three modifiable risk factors that have a more significant impact on diabetes-related complications than BMI are:
   a. Blood pressure, cholesterol, blood sugar levels
   b. Cortisol, HbA1c, blood sugar levels
   c. HbA1c, systolic blood pressure, cholesterol
   d. Urinary glucose, cholecystokinin, VO2max

4. The link between higher body weight and type 2 diabetes is most accurately referred to as:
   a. Correlation
   b. Causation
   c. Attribution
   d. Distribution

5. Which of the following lifestyle behaviours has been linked with a reduced risk of developing Type 2 diabetes according to the article by Jill Weisenberger?:
   a. Regularly consuming coffee and tea
   b. Getting 7-8 hours sleep most nights
   c. Lifting weights regularly
   d. Regularly consuming yogurt
   e. All of the above

Quiz Answers

Episode 1

1. = d., 2. = c., 3. = d., 4. = b., 5. = d.

Episode 2

1. = d., 2. = d., 3. = b., 4. = e., 5. = a.

Episode 3

1. = a., 2. = c., 3. = d., 4. = b., 5. = c.

Episode 4

1. = a., 2. = b., 3. = d., 4. = c., 5. = d.

Episode 5

1. = e., 2. = c., 3. = e., 4. = b., 5. = d.

Episode 6

1. = c., 2. = d., 3. = e., 4. = d., 5. = c.

Episode 7

1. = c., 2. = a., 3. = e., 4. = d., 5. = a.

Episode 8

1. = b., 2. = a., 3. = b., 4. = c., 5. = b.

Episode 9

1. = d., 2. = c., 3. = b., 4. = a., 5. = e.

Episode 10

1. = a., 2. = c., 3. = d., 4. = b., 5. = a.

Episode 11

1. = c., 2. = d., 3. = a., 4. = d., 5. = e.

Episode 12

1. = b., 2. = d., 3. = c., 4. = a., 5. = e.

www.ingramcontent.com/pod-product-compliance
Lightning Source LLC
Chambersburg PA
CBHW081100180526
45170CB00005B/1836